HANDBOOK OF PHYTOPHARMACOLOGY

(With Scientific Monograph of *Hypericum perforatum* L.)

Amritpal Singh Saroya
Herbal Consultant

SCIENTIFIC PUBLISHERS

Published by:
Pawan Kumar
Scientific Publishers (India)
5-A, New Pali Road, P.O. Box 91
Jodhpur - 342 001 (Raj.)
Tel.: +91-291-2433323
Fax.: + 91-291-2613449

E-mail: info@scientificpub.com
www.scientificpub.com

Print 2026

ISBN: 978-81-7233-552-6

eISBN: 978-93-8791-312-7

Cover design: [illegible] Sharma
Laser typeset: Rajesh Ojha
Printed in India

Contents

Part II
(*Hypericum perforatum* L.)

ALKANNA

Botanical Name: *Alkanna tinctoria* (L.) Taush. Syn *Alkanna tuberculata* (Forssk.) Meikle. Syn *Anchusa tinctoria* (L.) L.

Natural Order: Boraginaceae

Active Constituents: Napthoquinone: Alkannin (Leo *et al* 1992; Mita *et al* 1994; Gerardi *et al* 1998) and pyrrolizidine alkaloids: 7-angeloylretronecine, triangularine and dihydroxytriangularine.

Non-Clinical Studies: Antimicrobial, antioxidant (Assimopoulou, Papageorgiou and Boskou 2004).

Clinical Studies

Ulcers associated with varicose veins

The efficacy of ointment prepared from root extract of *Alkanna tinctoria* in treating venous ulcers was evaluated in seventy-two patients. The double-blind studies continued for three years. In majority of the patients, ulcers healed after thirty-five to forty-two days.

Toxicity: Hepatotoxic (due to presence of pyrrozolidine alkaloids).

BIBLIOGRAPHY

Assimopoulou, A.N.; Papageorgiou, V.P. and Boskou, D. (2004). Antioxidant activity of alkannin, shikonin and *Alkanna tinctoria* root extracts in oil substrates, *Food Chemistry* 87: 433-438.

Leo, P. De. *et al.* (1992). Extraction and purification of Alkannin from *Alkanna tinctoria.* Its location in the plant and time course accumulation in the tissues. *Agricoltura Mediterranea* 122: 334-339.

Mita, G. *et al.* (1994). Pigment production from *in vitro* cultures of *Alkanna tinctoria* Tausch. *Plant Cell Reports,* 13: 406-410.

Gerardi, C. *et al.* (1998). *Alkanna tinctoria*: *in vitro* culture and production of alkannin and other secondary metabolites. "Biotechnology in Agriculture and Forestry, Medicinal and Aromatic Plants" 10: 14-27 (Ed. Y.P.S. Bajai) Springer-Verlag.

AMALAKI

Botanical Name: *Phyllanthus emblica* Linn.

Natural Order: Euphorbiaceae.

Active Ingredients: Ascorbic acid or vitamin C (Giri 1939; Srinivasan 1944; Hanif *et al* 1966; Bramcahri and Gupta 1958; Deb and Chandrasekhara 1960; Soman and Pillay 1962), phyllembic acid (Theresa *et al* 1965), phyllembin or ethyl gallate (Jain 1974), procyanidin (Theresa *et al* 1965), amlaic acid (Theresa *et al* 1967), leucodelphinidin (Theresa *et al* 1965), 3-0-gallated prodelphidin (Theresa *et al* 1965), pyrogallol, ellagic acid (Thersa *et al* 1965), lupeol, corialgin (Thersa *et al* 1965; Zhang *et al* 2003), terchebin (Thersa *et al* 1965) emblicanin A and B, pedunculagin, punigluconin (Ghosal *et al* 2001), alkaloids: phyllantidine and phylanthine (Khana and Bansal 1975), flavonoid (rutin), putranjivain A, digallic acid (EL-Mekkawy *et al* 1995), emblicol, isostrictiniin (Zhang *et al* 2003), chebulinic acid (Zhang *et al* 2003) and chebulagic acid (Zhang *et al* 2003).

Pre Clinical Studies: Antibacterial (Unander 1992; Sankara-narayanan and Jolly 1993), anti-inflammatory (Kroes *et al* 1992; Asmawi *et al* 1992; Aswami and Moilanen 1977), anticancer (Chang *et al* 1985; Dhir *et al* 1991; Jose *et al* 1997; Jose, Joy and Kuttan 1999), antimutagenic (Jose *et al* 1997), antioxidant (Aruoma Murchia Butler and Halliwell 1993; Jose and Kuttan 1995), anti-stress (Bhattacharya,Ghosal and Bhattacharya 2000), antitussive (Nosal'ova Mokry and Hassan 2003), anti-proliferative (Khan *et al* 2002), modifier of nickel clastogenicity (Dhir *et al* 1991), cardio protective (Mand *et al* 1991; Mathur *et al* 1996), hepatoprotective (Gulati, Agarwal and Agarwal 1995), immuno-modulator (Suresh and Vasudevan 1994), hypolipidemic (Thakur 1985), hypotensive

(Ishaq *et al* 2005), acute pancreatitis (Throat *et al* 1995), gastroprotective (Al-Rehaily *et al* 2002), tardive dyskinesia (Bhattacharya Bhattacharya and Murugnandam 2000), sympathomimetic, antispasmodic and inhibiter of aldose reductase (Suryanarayana *et al* 2004).

Clinical Studies

Hypercholsterolemia

The effect on total serum cholesterol and its lipoprotein fractions of supplementation of the diet with *Emblica officinalis* was studied in normal and hypercholesterolaemic men aged 35-55 years. The supplement was given for a period of 28 days in the raw form. Both normal and hypercholesterolaemic subjects showed a decrease in cholesterol levels. Two weeks after withdrawing the supplement, the total serum cholesterol levels of the hypercholest-erolaemic subjects rose significantly almost to initial levels (Jacob *et al* 1988).

Dyspepsia

A study has been done to evaluate efficacy of *Emblica officinalis* in the treatment of dyspepsia (Chawla *et al* 1987).

Hyperacidity

A study was done in twenty patients to evaluate the efficacy of *Emblica officinalis* in hyperacidity (amlapitta in Ayurveda). The drug was given P.O. in dose of three G, thrice-a-day for seven days. The drug was found to be effective in eighty-five percent cases. However the drug failed to evoke response in hyperchlo-rhydria (Singh and Sharma 1971).

Kinetics

The kinetics of ascorbic acid degradation in *Emblica officinalis* as well as in pure ascorbic acid solutions at initial concentrations present in *Emblica officinalis* over a temperature range of 50-120°C (steady-state temperature) has been studied (Nisha *et al* 2004).

Toxicity

Acute Toxicity

Phyllemblin was well tolerated by mice in doses up to 100 mg/kg when given intraperitoneally in mice and upto 500 mg/kg when given orally. At doses above 500 mg/kg animals appeared drowsy and dull soon after the injection and were active again after 1-3 hours.

BIBLIOGRAPHY

Al-Rehaily, A.J., *et al.* (2002). Gastro protective effects of Amla (*Emblica officinalis*) on *in vivo* test models in rats. *Phyto-medicine* 9(6): 515-22.

Aruoma, O.I., Murchia, A., Butler, J. and Halliwell (1993). Evaluation of the antioxidant and prooxidant actions of gallic acid and its derivatives. *J. Agric. Food Chem* 41: 1880–1885.

Asmawi, M.Z. *et al.* (1992). Anti-inflammatory activities of *Emblica officinalis* Gaertm leaf extracts. *J. Pharm. Pharmacol* 45: 581–584.

Aswami, M.Z. and Moilanen, E. (1997). Anti-inflammatory activity of extracts from leaves of *Phyllanthus emblica. Planta Med* 63: 518–524.

Basa, S.C. and Shrinivasulu, C. (1987). Constituents of leaves of *Phyllanthus emblica* Linn. *Indian J. Nat. Prod* 3: 13–14.

Bhattacharya, A., Ghosal, S. and Bhattacharya, S.K. (2000). Antioxidant activity of tannoid principles of *Emblica officinalis* (amla) in chronic stress induced changes in rat brain Ind. *J. Exp. Biol.* 38 (9): 877-880.

Bhattacharya, S.K., Bhattacharya, D. and Murugnandam, A.V. (2000). Effect of *Emblica officinalis* tannoids on rat model of tardive dyskinesia, *Indian J. Exp. Biol.* 38 (9): 945.

Chang R.L. *et al.* (1985). Effect of ellagic acid and hydroxylated flavonoids on the tumourigenicity of benzo (a) pyrene and (+)-7ß, 8alfa-dihydro-9 alfa, 10 alfa-epoxy-7, 8, 9, 10-tetrahydrobenzo (a) pyrene on mouse skin in newborn mouse. *Carcinogenesis* 6: 1127–1133.

Chawla, Y.K. *et al.* (1987). Treatment of dyspepsia with amalaki *(Emblica officinalis),* an Ayurvedic drug, *Vagbhata* 5 (3): 24-26.

Dhir, H. *et al.* (1991). Modifying role of *Phyllanthus emblica* and ascorbic acid against nickel clastogenicity in mice. *Cancer Lett* 59: 9-18.

Dhir, R.H., Roy, A.K., Sharma, A. and Talukder, G. (1990). Protection afforded by aqueous extracts of *Phyllanthus* species against cyto toxicity induced by lead and aluminium salts. *Phytother. Res* 4: 172-176.

Gulati, R.K., Agarwal, S. and Agarwal, S.S. (1995).Hepatoprotective studies on *Phyllanthus emblica* Linn. and quercetin. *Ind. J. Exp. Biol* 33: 261-268, erratum 33: preceding table of contents.

Ishaq, G.M. *et al.* (2005). Hypotensive effect of aqueous extract of *Emblica officinalis* on anaesthetized dogs. *J-K Practitioner* 12 (4): 213-215.

Jacob, A., *et al.* (1988). Effect of the Indian gooseberry (amla) on serum cholesterol levels in men aged 35-55 years. *Eur. J. Clin. Nutr* 42: 939-944.

Jose, J.K. and Kuttan, R. (1995). Antioxidant activity of *Emblica officinalis. J. Clin Biochem. Nutr.*, 19: 63–70.

Jose, J.K. *et al.* (1997). Antimutagenic and anticarcinogenic activity of *Emblica officinalis* Gaertn. *J. Clin. Biochem. Nutr,* 22: 171–176.

Jose, J.K., Joy, K.L. and Kuttan, N.R. (1999). Effect of *Emblica officinalis, Phyllanthus amarus* and *Picrorrhiza kurroa* on *N*-nitrosodiethylamine induced hepatocarcinogenesis. *Cancer Lett* 136: 11–16.

Khan M.T. *et al.* (2002). Identification of pyrogallol as an antiproliferative compound present in extracts from the medicinal plant *Emblica officinalis*: effects on *in vitro* cell growth of human tumor cell lines. *Int. J. Oncol* 21(1): 187-92.

Khana, P. and Bansal, R. (1975). Phyllantidine and phyllantine from *Emblica officinalis* Gaertn leaves, fruits, and *in vitro* tissue cultures, *Indian J. Exp. Biol* 13: 82–83.

Kroes, B.H., *et al.* (1992). Anti-inflammatory activity of gallic acid. *Planta Med* 58: 499–504.

Mand, J.K., *et al.* (1991). Effect of amla (*Emblica officinalis*) on the development of atherosclerosis on hypercholesterolemic rabbits. *J. Res. Edu. Ind. Med* 10: 1–7.

Mathur, R. *et al.* (1996). Hypolipidaemic effect of fruit juice of *Emblica officinalis* in cholesterol-fed rabbits. *J. Ethnopharmacol,* 50: 61–68.

Nisha, P. *et al.* (2004). A study on degradation kinetics of ascorbic acid in amla (*Phyllanthus emblica* L.) during cooking. *Int J Food Sci Nutr* 55 (5): 415-422.

Nosal'ova, G., Mokry, J. and Hassan, K.M. (2003). Antitussive activity of the fruit extract of *Emblica officinalis* Gaertn. (Euphorbiaceae) Amla. *Phytomedicine* 10 (6-7): 583-9.

Sankaranarayanan, J. and Jolly, C.I. (1993). Phytochemical, antibacterial and pharmacological investigations on *Momordica charantia*

Linn., *Emblica officinalis* Gaertn and *Curcuma longa* Linn. *Indian J. Sci,* 1: 6-13.

Singh, B.N. and Sharma, P.V. (1971). Effect of amalaki on amlapitta, *J Res Ind Med* 5 (2): 223-230.

Suresh, K. and Vasudevan, D.M. (1994). Augmentation of murine natural killer cell and antibody dependent cellular cytotoxicity activities of *Phyllanthus emblica,* a new immunomodulator. *J. Ethnopharmacol,* 44: 55–60.

Suryanarayana, P. *et al.* (2004). Inhibition of aldose reductase by tannoid principles of Amla: implications for the prevention of sugar cataract. *Mol. Vis.* 12 (10): 148-54.

Thakur, C.P. (1985). *Emblica officinalis* reduces serum, aortic and hepatic cholesterol in rabbits. *Experientia* 41: 423–424.

Theresa, Y.M. *et al.* (1965). Studies on biosynthesis of tannins in indigenous plants XII. Occurrence of different polyphenolics in amla (*Phyllanthus emblica* Linn). *Leather Science* 12: 327–328.

Theresa, Y.M. *et al.* (1967). Studies on biosynthesis of tannins in indigenous plants XIII. Occurrence of a new gallotannin amlaic acid in amla leaves (*Phyllanthus emblica*). *Leather Science* 14: 16–17.

Throat, S.P. *et al.* (1995). *Emblica officinalis*: a novel therapy for acute pancreatitis - an experimental study. *HPB Surgery* 9: 25–30.

Zhang, L.Z. *et al.* (2003). Studies on chemical constituents in fruits of Tibetan medicine *Phyllanthus emblica. Zhongguo Zhong Yao Za Zhi* 28 (10): 940-3.

AMERICAN SPINKENARD

Botanical Name: *Aralia racemosa* Linn.

Natural Order: Araliaceae

Active Constituents: Polyenes: falicarinol, falicarindiole, saponins.

Non-Clinical Studies: Antimicrobial (Ahn *et al* 1990).

BIBLIOGRAPHY

Ahn, Y.J. *et al.* (1990). Selective growth responses of human intestinal bacteria to Arailaceae extracts. *Microbial Ecol Health Disease* 3: 223-229.

AMRAGANDHIHARIDRA

Botanical Name: *Curcuma amada* Roxb.

Natural Order: Zingiberaceae

Active Ingredients: Essential oil (contains α-pinene, β-curcumene, camphor and tumerone).

Non-Clinical Studies: Anti-inflammatory, hypolipdemic, antifungal.

ANNUAL WORMWOOD

Botanical Name: *Artemisia annua* Linn.

Natural Order: Asteraceae

Active Constituents: Sesquiterpene lactone: artemisinin (Kohler *et al* 1997; Van Geldre *et al* 1997) and flavonoids (Zheng 1994).

Non-Clinical Studies: Cytotoxic (Zheng 1994).

Clinical Studies

Malaria

132 patients were randomly assigned to receive 5 grams of *Artemisia annua* herb/L (1 liter/day for 7 days and quinine sulphate tablets (500 mg quinine sulphate, three times daily for 7 days). There was quick resolution of parasitaemia and clinical symptoms following 7 days of treatment. The cure rates were 74% with Artemisia compared with 91% for quinine (Mueller *et al* 2004).

Pharmacokinetics: Artemisinin is absorbed rapidly from the tea preparations rather than from capsules. The maximum plasma concentration reaches after 30 minutes following intake (Rath *et al* 2004).

Toxicity: LD $_{50}$ for artemisinin in mice: 4228 mg/kg (oral).

BIBLIOGRAPHY

Kohler, M., *et al.* (1997). Extraction of artemisinin and artemisinic acid from *Artemisia annua* L. using supercritical carbon dioxide. *J. Chromatogr* 785 (1-2): 353-60

Mueller, M.S. *et al.* (2004). Randomized controlled trial of a traditional preparation of *Artemisia annua* L. (Annual Wormwood) in the treatment of malaria. *Trans R Soc Trop Med Hyg* 98: 318-21.

Rath K, *et al.* (2004). Pharmacokinetic study of artemisinin after oral intake of a traditional preparation of *Artemisia annua* L. (annual wormwood). *Am J Trop. Med Hyg* 70: 128-132.

Van Geldre, E., *et al.* (1997). State of the art of the production of the anti-malarial compound artemisinin in plants. *Plant Mol Biol* 33 (2): 199-209.

Zheng, G.Q. (1994). Cytotoxic terpenoids and flavonoids from *Artemisia annua*. Planta Med 60 (1): 54-57.

ANTAMOOL (INDIAN IPECAUCH)

Botanical Name: *Tylophora indica* (Burm.f.) Merr. Syn *Tylophora asthmatica* (Linn.f.)

Natural Order: Asclepidaceae.

Active Ingredients: Phenanthroindalizidine alkaloids; tylophorine and tylophornine (Ratnagiriswaran and Venkatachalam 1935; Govindachari *et al* 1961; Gopalakrishnan *et al* 1979; Gopalakrishnan *et al* 1980).

Non-Clinical Studies: Anti-inflammatory (Gopalakrishnan *et al* 1979; Gopalakrishnan *et al* 1980), anti-asthmatic (Udupa, Udupa and Guruswamy 1991) and antispasmodic.

Clinical Studies

Bronchial Asthma

A. In a study, efficacy of *Tylophora asthmatica* was investigated as an anti-asthmatic agent. 110 bronchial asthma patients received one *Tylophora asthmatica* leaf (150 mg of the leaf by weight) or comparable placebo to be chewed and swallowed daily in the early morning for six days. At the end of one week, 62% of the patients consuming the Tylophora reported moderate to complete relief of asthma symptoms compared to 28% in the placebo group. Moreover, when patients were switched from the placebo to the active group and *vice versa,* similar positive trends could be seen, with 50% of the Tylophora group and 11% of the placebo group reporting symptomatic relief. In a follow-up study, the alcoholic extract of crude *Tylophora asthmatica* leaves in 1 gram of glucose had comparable effects to that of chewing the crude leaf, with 56% of the patients reporting moderate to complete improvement in asthmatic symptoms compared to 32% in the placebo group (Shivpuri, Singhal and Parkash 1972).

B. In another clinical trial, 30 patients with a diagnosis of bronchial asthma for at least two years were assigned at random to one of two treatment groups consisting of 15 individuals each. One group received either 350 mg of Tylophora leaf powder or placebo daily in the first week. In comparison, a second group of asthmatics were given a similar amount of the leaf for seven days followed by an anti-asthmatic drug combination. Overall, results of the study showed the amount of oxygen in the lung increased in those using the leaf but decreased in those using the placebo. In addition, those taking the herb had a notable night time reduction in their symptoms of shortness of breath. Unfortunately, Tylophora did not fare as well in improving asthmatic symptoms when compared to the anti-asthmatic drug combination (Thiruvengadam *et al* 1978).

C. In a double-blind, placebo-controlled, crossover study of 195 individuals with asthma, participants showed significant improvement when given 40 mg of a *Tylophora asthmatica* (alcohol extract) daily for 6 days as compared to placebo. Surprisingly, the difference was even more marked months after use of the herb was stopped (Mathew and Shivpuri 1974).

D. A higher quality double-blind study that enrolled 135 individuals found no benefit from *Tylophora asthmatica* in asthma (Gore, Rao and Guruswamy 1980).

Toxicity: Emetic in large doses (Dikshith, Raizada and Mulchandani 1990).

BIBLIOGRAPHY

Dikshith, T.S.; Raizada, R.B. and Mulchandani, N.B. (1990). Toxicity of pure alkaloid of *Tylophora asthamatica* in male rat. *Indian J Exp Biol*: 28: 208-212.

Gopalakrishnan, C. *et al.* (1979). Pharmacological investigations of tylophorine, the major alkaloid of *Tylophora indica*. *Indian J Med Res*: 69: 513-20.

Gopalakrishnan, C. *et al.* (1980). Effect of tylophorine, a major alkaloid of *Tylophora indica*, on immumopathological and inflammatory reactions. *Ind J Med Res*: 71: 940-8.

Gore, K.V.; Rao, A.K. and Guruswamy, M.N. (1980). Physiological studies with *Tylophora asthmatica* in bronchial asthma. *Indian J Med Res*: 71: 144-8.

Govindachari, T.R. *et al.* (1961). Chemical examination of *Tylophora asthmatica* -V. Structure of tylophorinine. *Tetrahedron*. 14: 288-295.

Gupta, S. *et al.* (1979). *Tylophora indica* in bronchial asthma double blind study. *Indian J Med Res* 69: 981-989.

Mathew, K.K. and Shivpuri, D.N. (1974). Treatment of asthma with alkaloids of *Tylophora indica*: a double-blind study. *Aspects Allergy Appl Immunol* 7: 166-179-7.

Ratnagiriswaran, A.N. and Venkatachalam, K. (1935). The chemical examination of *Tylophora asthmatica* and isolation of the alkaloids tylophorine and tylophorinine. *Indian Journal of Medical Research*, 22 (3): 433-441.

Shivpuri, D.N.; Singhal, S.C. and Parkash, D. (1972). Treatment of asthma with an alcoholic extract of *Tylophora indica*: a cross-over, double-blind study. *Ann Allergy* 30: 407-12.

Thiruvengadam, K.V. *et al.* (1978). *Tylophora indica* in bronchial asthma: a controlled comparison with a standard anti-asthmatic drug. *J Indian Med Assoc* 71: 172-6.

Udupa, A.L.; Udupa, S.L. and Guruswamy, M.N. (1991). The possible site of anti-asthmatic action of *Tylophora asthmatica* on pituitary-adrenal axis in albino rats. *Planta Med* 57 (5): 409-13

ARJUNA

Botanical Name: *Terminalia arjuna* (Roxb.) Wt.and Arn.

Natural Order: Combretaceae

Active Ingredients: Triterpenoid glycosides: arjunoside I, arjunoside II, arjunoside III, arjunoside IV, arjunglucoside I, arjunglucoside II, arjunglucoside III, arjunetin, friedelin, arjunaphthanoloside, terminoside A, arjungenin, arjunic acid, arjunolic acid, arjunetoside, oleanolic acid (Tsuyuki *et al* 1979; Anjaneyulu and Rama Prasad 1983; Ali *et al* 2003; Ali, *et al* 2003, Singh *et al* 2002; Upadhyay *et al* 2001), dihydroxy-triterpene carboxylic acid: terminic acid (Anjaneyulu and Rama Prasad 1983), ellagic acid (Kaur Grover and Kumar 1997), flavone: arjunolone (Sharma *et al* 1982), sterol: β-sitosterol, tannin: casuarinin (Kuo *et al* 2005), cardiac glycoside: 16,17-dihydroneridienone 3-O-beta-D-glucopyr-anosyl-(1->6)-O-beta-D-galactopyranoside (Yadev and Rathore 2001) and elagitannin: arjunin (Kandil and Nassar 1998).

Non-Clinical Studies: Antianginal and cardio protective (Colabawala 1951; Dwivedi and Agarwal 1994; Anand 1994; Dwivedi and Jauhri 1997; Bhatia *et al* 1998) antiarrythmic (Dwivedi Avasthi and Mahajan 1994), contraceptive (Gupta *et al* 1989) anti-cancer (Kandll and Nassar 1998), antimutagenic (Kaur Grover and Kumar 1997), hypocholesterolemic (Ram *et al* 1997; Shala Udupa and Udupa (1997), antihepatotoxic (Sharma *et al* 1991), antifertility (Yadev Mathur and Gupta 1980), antioxidant and platelet aggregation prevention (Sumitra *et al* 2001; Ali *et al* 2003), antihypertensive (Srinivas *et al* 2003), antiproliferative (Kuo *et al* 2005), anti-genotoxic (Scassellati-Sforzolini *et al* 1999), nitric oxide inhibitor (Ali *et al* 2003; Ali *et al* 2003) and antiviral (Cheng Lin and Lin 2002).

Clinical Studies

Congestive cardiac failure

A. In a study the bark powder of *Terminalia arjuna* was tested in treating congestive cardiac failure. It was observed that it showed a marked improvement in over 40% of the cases. Congestive cardiac failure due to congenital anomaly of heart and valvular disease was also controlled 4 out of 9 cases of CCF as a result of chronic bronchitis were also relieved by the treatment. Symptomatic complaints of essential hypertension (giddiness, insomnia, lassitude, headache and a lack of concentration ability) were relieved by *Terminalia arjuna*. Coagulation, bleeding and prothrombin time were be reduced by oral administration of an aqueous suspension of the bark powder (Phadnaik and Parasar 1981).

B. In a study cardio protective activity of bark extract of *Terminalia arjuna* was evaluated in twelve patients with refractory congestive cardiac failure. The extract demonstrated significant cardio protective activity as compared to placebo (Bharani, Ganguly and Bhargava 1995).

Mitral regurgitation

40 patients with fresh Acute Myocrardial Iinfarction (AMI) showing Ischemic Mitral Regurgitation (IMR) were randomly divided into 2 groups of 20 each. They were given placebo or 500 mg of *Terminalia arjuna* in addition to anti-ischemic treatment. After 1 and 3 months of follow up, patients receiving adjuvant *Terminalia arjuna* showed significant decrease in IMR, improvement in E : A ratio. Echocardiographic phases of ventri cular filling are described as early (E) and late, or atrial (A) and considerable reduction in anginal frequency (Dwivedi *et al* 2005).

BIBLIOGRAPHY

Ali, A. *et al.* (2003). A novel naphthanol glycoside from *Terminalia arjuna* with antioxidant and nitric oxide inhibitory activities. *Pharmazie* 58 (12): 932-934.

Ali, A. *et al.* (2003). Terminoside A, a new triterpene glycoside from the bark of *Terminalia arjuna* inhibits nitric oxide production in murine macrophages. *J Asian Nat Prod Res* 5(2): 137-142.

Anand, V. (1994). Antianginal and cardio protective effects of *Terminalia arjuna, J. Assoc. Physicians India* 42(9): 757.

Anjaneyulu, A.S.R. and Rama Prasad, A.V. (1983). Chemical examination of roots of *Terminalia arjuna* – The structures of arjunoside III and arjunoside IV, two new triterpenoid glycosides. *Phytochemistry* 21(8): 2057-2060.

Anjaneyulu, A.S.R. and Rama Prasad, A.V. (1983). Structures of terminic acid, a dihyrroxytriterpene carboxylic acid from *Terminalia arjuna, Phytochemistry* 22(4): 993-998.

Arora, R.C. *et al.* (1995). Evaluation of CTI (cardio protective drug) in subjects of coronary artery disease, hypertension and diabetes mellitus. *Flora and Fauna* 1 (2): 203-205.

Bharani, A.; Ganguly, A. and Bhargava, K.D. (1995). Salutary effect of *Terminalia arjuna* in patients with severe refractory heart failure. *Int. J. Cardiol* 49 (3): 191-199.

Bhatia, J. *et al.* (1998). Effect of *Terminalia arjuna* on coronary flow-an experimental study. *Indian J Pharmacol* 30 (2): 118-119.

Cheng, H.Y.; Lin, C.C. and Lin, T.C. (2002). Antiherpes simplex virus type 2 activity of casuarinin from the bark of *Terminalia arjuna* Linn. *Antiviral Res* 55 (3): 447-455.

Colabawala, H.M. (1951). An evaluation of the cardiotonic and other properties of *Terminalia arjuna. Indian Heart J* 3: 205-230.

Dwivedi, S. and Jauhri, R. (1997). Beneficial effects of *Terminalia arjuna* in coronary artery disease. *Indian Heart J* 49 (5): 507-510.

Dwivedi, S. and Agarwal, M.P. (1994). Antianginal aand cardio protective effects of *Terminalia arjuna* in coronary artery disease. *J Assoc Physicians India* 42 (4): 287-289.

Dwivedi, S.; Avasthi, S. and Mahajan, S. (1994). Role of *Terminalia arjuna* in ventricular tachyarrythmias. *Indian Pract* 57 (6): 523-525.

Dwivedi, S. *et al.* (2005). Role of *Terminalia arjuna* in ischaemic mitral regurgitation. *Ind J Cardiol* 100 (3): 507-508.

Gupta, D.N. *et al.* (1989). Postcoital contraceptive efficacy of *Terminalia arjuna* on albino rats. *Fitoterapia* 60 (3): 275-276.

Kandil, F.E. and Nassar, M.I. (1998). A tannin anti-cancer promoter from *Terminalia arjuna. Phytochemistry* 47 (8): 1567-1568.

Kaur, S.; Grover, I.S. and Kumar, S. (1997). Antimutagenic potential of ellagic acid isolated from *Terminalia arjuna. Indian J Exp Biol* 35 (5): 478-482.

Kuo, P.L. *et al.* (2005). Casuarinin from the bark of *Terminalia arjuna* induces apoptosis and cell cycle arrest in human breast adeno-carcinoma MCF-7 cells. *Planta Med* 71 (3): 237-243.

Phadnaik, B.S. and Parasar, G.C. (1981). Haematological studies on *Terminalia arjuna* in respect of blood coagulation. *Punjabrao Krishi Vidyapeeth Res* J. 5: 1.

Ram, A. *et al.* (1997). Hypocholesterolemic effects of *Terminalia* arjuna tree bark. *J Ethnopharmacol* 55 (3): 165-169.

Scassellati-Sforzolini, G. *et al.* (1999). Antigenotoxic properties of *Terminalia arjuna* bark extracts. *J Environ Pathol Toxicol Oncol* 18 (2): 119-125.

Shala, H.P.; Udupa, S.L. and Udupa, A.L. (1997). Hyperlipidemic effect of *Terminalia arjuna* in cholesterol fed rabbits *Fitoterapia* 68(5):405-409.

Sharma, P.N. *et al.* (1982). Arjunolone-a new flavone from stem bark of *Terminalia arjuna. Indian J Chem* 21B (3): 263-264.

Sharma, S.S. *et al.* (1991). Antihepatotoxic activity of *Terminalia arjuna,* Proc. 24th Indian Pharmacol Soc Conf Ahmedabad, Gujarat India 29-31: 2.

Singh, D.V. *et al.* (2002). RP-LC determination of oleane derivatives in *Terminalia arjuna. J Pharm Biomed Anal* 28 (3-4): 447-452.

Srinivas, N. *et al.* (2003). Possible mechanisms of hypotension produced 70% alcoholic extract of *Terminalia arjuna* (L.) in anaesthetized dogs. *BMC Complement Altern Med* 3: 5.

Sumitra, M *et al.* (2001). Experimental myocardial necrosis in rats: role of arjunolic acid on platelet aggregation, coagulation and antioxidant status. *Mol Cell Biochem* 224 (1-2): 135-42.

Tsuyuki, T. *et al.* (1979). New triterpene glucoside from *Terminalia arjuna*, Arjunglucoside III. *Bull Chem Soc Japan* 52 (10): 3127-3128.

Upadhyay, R.K. *et al.* (2001). Triterpene glycoside from *Terminalia arjuna. J Asian Nat Prod Res* 3 (3): 207-212.

Yadev, B.L.; Mathur, R. and Gupta, R.B. (1980). Antifertility activity of *Terminalia arjuna* W. and A. in female albino rats. *Probe* 19 (3): 196-198.

Yadev, R.N. and Rathore, K. (2001). A new cardenolide from the roots of *Terminalia arjuna. Fitoterpia 72* (4): 459-461.

ARKA

Botanical Name: *Calotropis procera* (Ait.) Ait. f.

Natural Order: Asclepidaceae.

Active Ingredients: Cardiac glycosides: calotropine, principle of peculiar nature: mudarine, pentacylic triterpenes (Ansari and Ali, 2001), organic carbonate (Olea *et al* 2002), sterols: stigmasterol and beta-sitosterol (Olea *et al* 2002).

Non-Clinical Studies: Antimicrobial (Kumar and Arya 2006), antipyretic (Dewan, Kumar and Kumar 2000), antifungal (Tharani 2005), cardio protective (Ahmad Rana and Dixit (2004) cytostatic (Anon: PDR for Herbal Medicines; Kumar and Arya 2006), lipolytic (Akinloye *et al* 2001) antiseptic (Fleurentin and Pelt 1982) and anti-inflammatory (Singh *et al* 2000).

Adverse Drug Reaction: Bullous eruptions (Behl and Luthra 2002).

Toxicity: Cytolysis reported in kidney tubules of Wistar rats (Al-Robai *et al.,* 1993a). Toxic iridocyclitis and allergic contact dermatitis has been reported (Tomar, Agarwal and Agarwal 1970; Handa, Sadana and Sharma 1984).

BIBLIOGRAPHY

Ahmad, K.K.; Rana, A.C. and Dixit, V.K. (2004). Effect of *Calotropis procera* latex on isoproterenol induced myocardial infarction in albino rats. *Phytomedicine* 11(4): 327-330.

Akinloye, A.K. *et al.* (2001). Lipolytic effect of *Calotropis procera* on skin of wistar rats. *African Journal of Biomedical Research*, 4 (3): 143- 145.

Al-Robai *et al.* (1993a). Toxicological studies on the latex of the usher plants *Calotropis procera* (Alt). R.Br. In Saudi Arabia. III Effects of usher latex on the fine structures, oxygen consumption and Na+'K+- Transporting At Pase Activity of Albino Rat kidneys. *Arab Gulf Journal of Scientific Research* 11(3): 441-455.

Anon: PDR for Herbal Medicines (2000). Medical Economics Company Inc, Montvale NJ: 812-813.

Ansari, S.H. and Ali, M. (2001). Norditerpenic ester and pentacyclic triterpenoids from root bark of *Calotropis procera* (Ait.) R. Br. *Pharmazie* 56(2): 175-177.

Behl, P.N. and Luthra, A. (2002). Bullous eruption with *Calotropis procera* - A medicinal plant used in India. *Indian Journal of Dermatology, Venerology and Leprology* 68(3): 150-151.

Dewan, S.; Kumar, S. and Kumar, V.L. (2000). Antipyretic effect of latex of *Calotropis procera. Indian Journal of Pharmacology* 32(3): 252.

Fleurentin, J. and and Pelt, J.M. (1982). Repertory of drugs and Medicinal plants of Yemen. *J. Ethnopharmacol* 5: 85-108.

Handa, F.; Sadana, J.K. and Sharma, P.K. (1984). Allergic contact dermatitis due to plant of *Calotropis procera. Indian J Dermatol* 29: 27-29.

Kumar, V.L. and Arya, S. (2006). Medicinal uses and Pharmacological properties of *Calotropis procera*. Rec Prog Med Plan.

Olea, R.S. *et al.* (2002). Organic carbonate from *Calotropis procera* leaves *Fitoterapia* 73 (3): 263-265.

Singh, H. *et al.* (2000). Inflammation induce by latex of *Calotropis procera* a new model to evaluate anti-inflammatory drugs. *J Pharmacol Toxicol Methods* 43: 219-224

Tharani, C.B. (2005). Inhibitory effects of latex of *Calotropis procera* against *Candida albicans. Indian Journal of Pharmacology* 37(5): 334-335.

Tomar, V.P.S.; Agarwal, P.K. and Agarwal, B.L. (1970). Toxic irido-cyclitis caused by *Calotropis procera. J All-India Ophthalmol Soc* 18: 15-16.

ARNICA

Botanical Name: *Arnica montana* Linn.

Natural Order: Asteraceae

Active Constituents: Sesquiterpene lactone: arnifolin, helenalin, 11, 13-dihydroxyhelenalin (Schroder *et al* 1990; Schmidt *et al* 1998), polyactyelenes, hydroxycoumarins, flavonoids, bitter glucoside: arnicin, sterol: arnisterol and carbohydrates: inulin and phulin.

Non-Clinical Studies: Antimicrobial (Koo *et al* 2000), anti-inflammatory (Lyss *et al* 1997), anti-platelet (Schroder *et al* 1990) and peripheral vasodilator.

Clinical Studies

Osteoarthritis

An open multicenter trial, investigated the safety and efficacy of an *Arnica montana* fresh plant gel, applied twice daily, in 26 men and 53 women with mild to moderate osteoarthritis of the knee. After 3 and 6 weeks, significant decreases in median total scores on the Western Ontario and McMaster Universities Osteoarthritis Index (WOMAC) were evident in the intention-to-treat and per-protocol populations (both $P < 0.0001$). Scores on the pain, stiffness and function subscales also showed significant reductions at these timepoints. The overall local adverse-event rate of 7.6% included only one allergic reaction. Sixty-nine patients (87%) rated the tolerability of the gel as "good" or "fairly good" and 76% would use it again. Topical application of *Arnica montana* gel for 6 weeks was a safe, well-tolerated, and effective treatment of mild to moderate osteoarthritis of the knee (Knuesel *et al* 2002).

Muscle soreness

A randomized, double-blind, placebo-controlled trial reported ineffectiveness of Arnica tincture in treating muscle soreness (Vickers *et al* 1998).

Adverse Drug Reations: Allergic dermatitis (Hormann and Kortin, 1995).

BIBLIOGRAPHY

Hormann, H.P. and Kortin, H.C. (1995). Allergic contact dermatitis due to Arnica tincture self-medication. *Phytomedicine* 4: 315-317.

Knuesel O., *et al.* (2002). *Arnica montana* gel in osteoarthritis of the knee: an open, multicenter clinical trial. *Adv Ther* 19(5): 209-18.

Koo, H. *et al.* (2000). *In vitro* antimicrobial activity of propolis and *Arnica montana* against oral pathogens. *Arch Oral Biol* 45(2): 141-8.

Lyss, G.; *et al.* (1997). Helenalin, an anti-inflammatory sesquiterpene lactone from *Arnica*, selectively inhibits transcription factor NF-kappa B *Biol Chem* 378 (9): 951-961.

Schmidt, T.J. *et al.* (1998). Sesquiterpene lactone content in leaves of *in vitro* and field cultivated *Arnica montana*. *Planta Med* 64 (3): 268-270.

Schroder, H. *et al.* (1990). Helenalin and 11 alpha, 13-dihydrohe-lenalin, two constituents from *Arnica montana* L., inhibit human platelet function via thiol-dependent pathways. *Thromb Res* 57: 839-845.

Vickers, A.J. *et al.* (1998). Homeopathic *Arnica* 30X is ineffective for muscle soreness after long-distance running: a randomized, double-blind, placebo-controlled trial. *Clin J Pain* 14 (3): 227-231.

Wijnsma, R; Woerdenbag, H.J. and Busse, W. (1995). The importance of *Arnica*-species in phytomedicine. *Z Phytother* 16 (1): 48-62.

ASHWAGANDHA

Botanical Name: *Withania somnifera* (Linn.) Dunal

Natural Order: Solanaceae

Active Ingredients: Alkaloids: anahygrine (Srivastava, Iyer and Ray 1960; Schwarting *et al* 1963; Leary *et al* 1964), anaferine (Rother *et al* 1962; Schwarting *et al* 1963), cuscohygrine (Schwarting *et al* 1963), hygrine, isopelletriene (Schwarting *et al* 1963), nicotine (Atal *et al* 1975), pseudopelletriene (Schwarting *et al* 1963) pseudowithanine, 3-togloxytropine, 3-tropyltigloate, tropine (Schwarting *et al* 1963), somniferine (Atal *et al* 1975), withasomnine (Schroter and Neumann 1966; Kulinkovich Masalov *et al* 1996; Morimoto *et al*), withanine (Atal *et al* 1975), withaninine (Atal *et al* 1975), withanone (Khana *et al* 1962), visamine (Truntneva 1971; Polyiet *et al.*, 1969), withanolides: withaferin-A (Bahr. and Hansel 1962; Choudhary and Neogy 1975; Devi *et al* 1992; Devi 1996; Uma *et al* 1996; Gupta *et al* 1996), 37–deoxy-14 a-hydroxy withaferin A (Glotter Waitman and Lavie 1966), withaniol, 1α, 3B, 20-trihydroxy witha-5, 24-dienolide (Velde and Lavie), 7α, 27-hydroxy-1-oxo-witha-2, 5-24-trienolide (Kirson *et al* 1971), withanolide C (Bessalle and Lavie 1992), withanolide D (Choudharyand and Neogy 1975), withanolide E (Glotter Abraham and Guenzberg 1977), withanolide Q (Kirson, Glotter and Abraham 1975; Das *et al* 1985), withanolide R (Kirson, Glotter and Abraham 1975), withanolide S (Glotter, Kirson and Abharam 1971), withanolide Y (Bessele Lavie and Frolow 1987; Bessalle and Lavie 1992), withasomnilide (Ali, Shuab and Ansari 1977), withasomniferabolide (Ali, Shuab, Ansari 1977), somniferanolide (Ali, Shuab and Ansari 1977), somnwithanolide (Ali, Shuab and Ansari 1977), withasomidienone (1993 Shahwar), dehydro-withanilide-R, withasomniferin-A, withasomniferol A-C, withstramonolide or 12 deoxy- withstramonolide viscosalac-tone B, 2-3-dihydrowithaferin-A, 24, 25-dihydrowithaferin-A, physagulin D (Jayaprakasam *et al* 2003), 27-O-glucosides: sitoinoside VII,

VIII, IX and X (Ghosal, Kaur and Srivastava 1988), coagulin G, coagulin F, withaniol, withaferinil, withacoagin, withanone (Ali *et al* 1997), glycosides : withaosides 1- VII (Mastuda *et al* 2001), flavonol glycosides and phenolics: dihydroxykaempferol 3-rutinoside, quercetin and its 3-O-rutinoside and 3-rutinoside-7-glucoside (Kandil *et al* 1994).

Non-Clinical Studies: Antibacterial (Das, 1964; Jaffer, Jawad *et al* 1988), antidepressant (Mehta *et al* 1991), anti-inflammatory (Begum and Sadique 1987), anti-tumor (Shohat *et al* 1976; Chatterjee and Chakraborti 1980; Das *et al* 1985; al-Khafaji and Abdul-Nabi 1992; Devi *et al* 1992; Devi 1996; Sharada *et al* 1996; Sharma *et al* 1998; Jayaprakasam *et al* 2003; Padmavathi, *et al* 2005), immunomodulator (Ziauddin *et al* 1996; Davis and Kuttan 2000), antioxidant (Bhattacharya, Ghosal and Bhattacharya 2001), immunosupressant (Shohat, Kirson and Lavie 1978), anti-arthritic (Rasool *et al* 2000), anti-stress or adaptogen (Grandhi, Mujumdar and Patwardhan 1994; Dhuley 2000), hypoglycemic, diuretic, hepatoprotective (Sudhir and Budhiraja 1992), hypocholesterolemic, reno protectie (Davis and Kuttan 2000), cardioprotective (Dhuley 2000), nervine (Kuboyama *et al* 2000), antimasculinic (Rtansooriya and Weersooriya 2002), anticholinesterase (Houghton Shipman and Howes 2000), reversal of paclitaxel induced neutropenia (Gupta *et al* 2001), nootropic (Dhuley 2001), angiogenesis inhibitor (Mohan *et al* 2004), Inhibitory action against tachyphylaxis to clonidine (Mastuda *et al* 2001), analgesic (Sakka, Mazen, Pavelescu and Grigorescu 1990), hypotensive (Malhotra, Das and Dhalla 1960).

Clinical Studies

Anxiety neurosis

The effect of Ashwagandharishta (medicated wine prepared from the roots of *Withania somnifera*) was studied in 30 patients with anxiety neurosis. Moderate improvement in palpitation, tremors, headache, anorexia, lack of concentration, dyspepsia, fatigue and irritability was observed, while maximum improvement was seen in nervousness with 40 ml of Ashwagandharishta administered in two divided doses for one month. No side effects were observed with Ashwagandharishta.

Rheumatoid arthritis

Root powder of *Withania Sominifera* was studied in 46 patients of rheumatoid arthritis with a dose of 4, 6 or 9 gm/day for a period of 3 to 4 weeks. Pain and swelling disappeared completely in 14 patients; considerable improvement was observed in 10 patients and 11 patients showed mild improvement. There was no relief in 4 patients and 7 patients discontinued the treatments. Root powder in a dose of 9 gm/day in divided doses used for 3 to 4 weeks in patients with arthritis was well tolerated.

Toxicity

A 180-day study involving rats found unfavorable increases in catecholamine content of the heart and decreases in the adrenal glands (Sharada, Emerson and Devi 1993). The LD_{50} of leaf extract was found to be > 1000 mg kg^{-1}, i.p. in mice (Review of Natural Products). Acute toxicity 465/Kg in rats; 432 mg/Kg in mice (Malhotra *et al* 1965), alcoholic extract: 1750+/- 41mg p.o. (Singh *et al* 1982), aqueous-methanol extract: WS (SG-1): 1076=/- 78 mg kg i.p. WS (SG-2; equimolar proportions of sitoinosides V11 and V111 and withaferin-A): 1564 +/- 92 mg/Kg i.p. (Grandhi Majumdar and Pathwardhan 1994), 50% alcoholic extract: 1000 mg/kg in rats (Aphale *et al* 1998); LD for visamine: 450mg/kg s.c and 125 mg/kg i.p. in mice. Anagyrine is teratogenic (Keeler, 1989).

BIBLIOGRAPHY

Al-Hindawi M.K.; al-Khafaji, S.H. and Abdul-Nabi, M.H. (1992). Anti-granuloma activity of Iraqi *Withania somnifera*. *J Ethnopharmacol* 37 (2): 113-6.

Ali, M., Shuaib, M. *et al.* (1997). Withanolides from the stem bark of *Withania somnifera*. *Phytochemistry* 44 (6): 1163-1168.

Aphale, A.A. *et al.* (1998). *Indian J Physiol,. Pharmacol*; 42 (2): 299-302.

Bahr, V. and Hansel, R. (1962). *Planta Medica* 44 (1): 32-33.

Begum, V.H. and Sadique, J. (1987). Effect of *Withania somnifera* on glycosaminoglycan synthesis in carrageenin-induced air pouch granuloma. *Biochem Med Metab Biol* 38 (3): 272-7.

Bessalle, R and Lavie, D. (1992). *Phytochemistry* 31 (10): 3648-3651.

Bessele R., Lavie D. and Frolow F. (1987). Withanaloid Y, a withanolide from a hybrid *Withania somnifera. Phytochemistry* 26: 1797-1800.

Bhattacharya, A.; Ghosal, S. and Bhattacharya, S.K. (1995). Effects of glycowithanolides from *Withania somnifera* on an animal model of Alzheimer's disease and perturbed central cholinergic markers of cognition in rats. *Phytother Res* 9: 110–3.

Bhattacharya, A.; Ghosal, S. and and Bhattacharya, S.K. (2001). Antioxidant effect of *Withania somnifera* glycowithanolides in chronic footshock stress-induced perturbations of oxidative free radical scavenging enzymes and lipid peroxidation in rat frontal and striatum. *J Ethnopharmacol* 74 (1): 1-6.

Chatterjee, S. and Chakraborti, S.K. (1980). Antimicrobial activities of some antineoplastic and other withanolides. *Antonie Van Leeuwenhoek* 46 (1): 59-63.

Choudhary, K. and Neogy, R.K. (1975). *Biochem. Pharmacol* 24(8): 919-920.; *Biol Abstr* 60, 62, 270.

Das, H. *et al.* (1985). Anti-neoplastic agents from plants. Antitumour activity of Withanolide D. *Indian J Cancer Chemother* 7: 59-65.

Davis, L. and Kuttan, G. (2000). Immunomodulatory activity of *Withania somnifera. J Ethnopharmacol* 7 (1-2): 193-200.

Davis, L. and G. Kuttan (2000). Effect of *Withania somnifera* on cyclophosphamide-induced urotoxicity. *Cancer Letters* 148 (1): 9-17.

Devi PU *et al.* (1992). *In vivo* growth inhibitory effect of *Withania somnifera* (Ashwagandha) on a transplantable mouse tumor, Sarcoma 180. *Indian J Exp Biol* 30 (3): 169-172.

Devi PU. (1996). *Withania somnifera* Dunal (Ashwagandha): potential plant source of a promising drug for cancer chemotherapy and radiosensitization. *Indian J Exp Biol* 34: 927-932.

Dhuley, J.N. (2000). Adaptogenic and cardioprotective action of ashwagandha in rats and frogs. *Journal of Ethnopharmacology* 70 (1): 57-63.

Dhuley, J.N. (2001). *Phytother Res* 15 (6): 524-528.

Elaskka, M. *et al.* (1990). New data referring to chemistry of *Withania somnifera* species. *Rev Med Chir Soc Med Nat Iasi* 94 (2): 385-387.

Ghosal, S., Lal J. and Srivastava R.S. (1988). Immunomodulatory and CNS effect of sitoindosides IX and X, two regio glycowithanolides from *W somnifera. Indian J Nat Prod* 4: 12-13.

Glotter, E., Abraham, A. and Guenzberg, I.K. (1977). Naturally occurring steroidal lactones with 17 a-oriented side chain. Structure of Withanolide E and related compounds. *J Chem Soc Perkins Trans-* 1: 341-43.

Glotter, E., Waitman, R. and Lavie, D. (1966). Constitutents of Wtihania somnifera VIII a new steroidal lactone 37-deoxy-14 a-hydroxy withaferin A. *J Chem Soc* 19: 1765-67

Grandhi, A.; Majumdar, A.M. and Pathwardhan, B. (1994). A comparative pharmacological investigation of Ashwagandha and Ginseng. *J Ethnopharmacol* 44: 131-135.

Gupta, Y.K. *et al.* (2001). Reversal of paclitaxel induced neutropenia by *Withania somnifera* in mice. *Indian J Physiol Pharmacol*, 45: 253-7.

Houghton, P.J.; Shipman, K. and Howes, M.J. (2000). Inhibitiion of cholinesterase by extracts and fractions of *Withania somnifera* root. *J Pharm Pharmacol* 52: (Suppl.) 273.

Illayperuma, I.; Rtansooriya, W.D. and Weersooriya, T.R. (2002). Effect of *Withania somnifera* root extract on the sexual behavior of male rats. *Asian Journal of Andrology* 4: 295-298.

Jaffer, H.J. *et al.* (1988). Evaluation of antimicrobial activity of *Withania somnifera* extracts. *Fitoterapia* (6): 497-500.

Jayaprakasam, B. *et al.* (2003). Growth inhibitor of human tumour cell lines by withanolides from *Withania somnifera* leaves. *Life Sci* 74: 125-132.

Kandil, F.E. *et al.* (1994). Flavonol glycosides and phenolics from *Withania somnifera. Phytochemistry* 37 (4): 1215-1216.

Keeler, R.F. (1989). Quinolizidine alkaloids in range and grain lupins. Pages 133-167 in Cheeke, P.R., ed. Toxicants of plant origin. Vol. I. Alkaloids. CRC Press, Inc., Boca Raton, Fla., USA. 335

Kulinkovich, O., *et al.* (1996). New synthetic route to the alkaloid whithasomnine by ring transformation of a functionalized cyclopropanol via the parent pyrrolo (1,2-b) pyrazole. *Tetrahedron Letters* 37 (7): 1095-1096.

Kulkarni, S.K. and Ninan, I. (1997). Inhibition of morphine tolerance and dependence by *Withania somnifera* in mice. *J Ethno Pharmacology* 57(3): 213-17.

Leary, J.D. *et al.* (1964) *Chem Abstr* 60,107836 g.

Malhotra, C.L. *et al.* (1965). The effect of total alkaloids on the smooth muscles. *Indian J Physiol Pharmacol* 9: 9-15.

Malhotra, C.L., Das, P.K. and Dhalla, N.S. (1960). Studies of *Withania ashawagandha*. II Effect of total extracts on cardiovascular system, respiration, Skeletal muscles. *Indian J Physiol Pharmacol* 4: 46-49.

Mastuda, H. *et al.* (2001). Inhibitory action of withanosides for tachyphylaxis to clonidine in isolated guinea-pig ileum. *Bioorganic and Medicinal Chemistry* 9 (6): 1499-1507 (9).

Mehta, A.K. *et al.* (1991). Pharmacological effects of *Withania somnifera* root extract on GABA receptor complex. In: *Indian J Med Res* 94: 312-315.

Mohan, R. *et al.* (2004). Withaferin-A potentially inhibits angiogenesis and NF-K B activation by targeting the ubiquitin-mediated proteasome pathway. *Angiogenesis* 7, 115.

Ozguven, M. *et al.* (1991). Drug yields and withanolide contents of *Withania somnifera* Dunal cultivated in Turkiye. *Gazi Universitesi Eczacilik Fakultesi Dergisi* 8 (2): 89-100.

Padmavathi, B. *et al.* (2005). Roots of *Withania somnifera* inhibit fore stomach and skin carcinogenesis in mice. *eCAM* 2 (1): 99-105.

Rasool, M., *et al.* (2000). Effect of *Withania somnifera* on lysosomal acid hydrolases in adjuvant-induced arthritis in rats. *Pharmacy and Pharmacology Communications* 6 (4): 187-190.

Rother, A., *et al.* (1962). *Chem Ind* No.14 654-655.

Sakka Mazen, E., Pavelescu, M. and Grigorescu, E. (1990). Contributions to the pharmacodynamic study of roots of *Withania somnifera* Dun species of Pakistani origin. Note III: Testing of analgesic activity of dichlormethanic and methanolic extract from *Withania somnifera* roots. *Rev Med Chir Soc Med Nat Iasi* 94 (3-4): 603-05.

Schwarting, A.E. *et al.* (1963). The alkaloids of *W somnifera*. *Llyoida* 26: 258-73.

Sharada, A.C. *et al.* (1996). Antitumor and radiosensitizing effects of withaferin A on mouse Ehrlich ascites carcinoma *in vivo*. *Acta Oncologica Stockholm* 35(1): 95-100.

Sharada, A.C., Emerson, S.F. and Uma Devi, P. (1993). Toxicity of *Withania somnifera* root extracts in rats and mice. *International Journal of Pharmacognosy*. 31: 205-12.

Sharma, I.P., Singh, V.V., Kamath, R., Uma Devi P. and Udupa, N. (1998). Niosomal Withaferin A with better antitumour efficacy. *Indian J Pharmaceutical Sciences* 60 (4): 45-48.

Shohat, B., Kirson, I. and Lavie, D. (1978). Immuno depressive properties of withaferin and withanolide D. *Biomedicines* 28: 18-23.

Shohat, *et al.* (1976). The effect of Withaferin-A. on human peripheral blood lymphocytes. An electro microscope study. *Cancer Lett* 2(2):63-70.

Singh, N. *et al.* (1982). *Int. J. Crude Drug Res*; 20; 29-35.

Srivastava, S.K., Iyer, S.S. and Ray, G.K. (1960). Estimation of the total alkaloids of *Withania somnifera* (Dunal). *Indian Journal of Pharmacy* 22 (4): 94-95.

Sudhir, S. and Budhiraja, R.D. (1992). Comparison of the protective effect of withaferin 'A' and hydrocortisone against carbon tetrachloride induced hepatotoxicity in rats. *Indian Journal of Physiology and Pharmacology* 36 (2): 127-129.

Tohda, C., *et al.* (2000). Dendrite extension by methanol extract of Ashwagandha (roots of *Withania somnifera*) in SK-N-SH cells. Neuro report 11 (9): 1981-1985.

Truntneva, E.A. (1971). *Chem. Abstr.*, 79, 273152.

Trutneva, E.A. (1971). *Pharmacology* of new alkaloid Visamine (causes prolonged hexenal incuded sleeping time and have hypothermic and nicotinolytic effects) from *Withania somnifera. Tr. Vses, Nauch-lssled Inst Lek Rast* 14: 159-62.

Uma, D.P. *et al.* (1996). *International Journal of Radiation Biology* 69 (2): 193-197.

Withania somnifera. Review of Natural Products. St. Louis: Facts and Comparisons. Oct. 2000.

Ziauddin, M., *et al.* (1996), Studies on the immunomodulatory effects of Ashwagandha. *J Ethnopharmacol* 50: 69-76.

BALA (COUNTRY MALLOW)

Botanical Name: *Sida cordifolia* Linn.

Natural Order: Malvaceae

Active Ingredients: Asparagin, alkaloids: hypaphorine, ephedrine, vasicine, phytosterols, mucin, gelatin and potassium nitrate, bioflavonoids: rutin.

Non-Clinical Studies: Analgesic, anti-inflammatory and hypoglycemic (Kanth and Diwan 1999; Franzoti *et al* 2000) and hepatoprotective.

Toxicity: Aqueous extract of *Sida cordifolia* has low toxicity (Franzoti *et al* 2000).

BIBLIOGRAPHY

Franzoti, E.M. *et al.* (2000). Anti-inflammatory, analgesic activity and acute toxicity of *Sida cordifolia* L. (Malva-branca). *J. Ethnopharmacology* 72 (1-2): 273-7.

Kanth, V.R. and Diwan, P.V. (1999). Analgesic, anti-inflammatory and hypoglycemic activities of *Sida cordifolia. Phytother Res* 13 (1): 75-7.

BHALLATAKA (MARKING NUT)

Botanical Name: *Semecarpus anacardium* Linn.

Natural Order: Anacardiaceae

Active Ingredients: Oil, phenolic compounds: bhilawanol, semecaprol.

Non-Clinical Studies: Hypolipidemic (Sharma, Mathur and Dixit 1995).

BIBLIOGRAPHY

Sharma, A.; Mathur, R. and Dixit, V.P. (1995). Hypercholesterolemic activity of nut shell extract of *Semacarpus anacardium* (Bhilawa) in cholesterol fed rabbits. *Indian J Exp Biol* 33: 444-8.

BHRINGRAJA (TRAILING ECLIPTA)

Botanical Name: *Eclipta alba* (L.) Hassk.

Natural Order: Asteraceae.

Active Ingredients: Resin, alkaloids: ecliptine, 25-ß-hydroxyverazine, coumarin: wedelolactone, demethylwedelolactone (Syed *et al* 2003), triterpenoid glucosides: eclabatin, ecliptasaponin C (Zhang and Chen 1996), ecliptasaponin D (Zhang *et al* 1997), wedelic acid, alpha-amyrin, ursolic acid, oleanolic acid, taraxstane glycosides (Yahara *et al* 1997).

Non-Clinical Studies: Analgesic (Swant, Issac and Narayanan 2004), heaptoprotective (Dixit and Achar 1979; Singh*et al*

2001; Saxena, Singh and Anand 1993) and trypsin inhibitory (Syed *et al* 2003).

BIBLIOGRAPHY

Dixit, S.P. and Achar, M.P. (1979). Bhringaraj in the treatment of infective hepatitis. *Curr Med Pract* 23 (60): 237-42.

Saxena, A.K.; Singh, B. and Anand, K.K. (1993). Hepatoprotective effects of *Eclipta alba* on sub cellular levels in rats. *J Ethnopharmacol* 40 (3): 155-61

Singh, B. *et al.* (2001). *In vivo* hepatoprotective activity of active fraction of ethanolic extract of *Eclipta alba* leaves. *Indian J Physiol Pharmacol* 45 (4): 435-41.

Swant, M.; Issac, J.C. and Narayanan, S. (2004). Analgesic studies on total alkaloids and alcohol extracts of *Eclipta alba* (Linn.) Hassk. *Phytother Res* 18 (2): 111-3.

Syed, S.D. *et al.* (2003). Trypsin inhibitory effect of wedelolactone and demethylwedelolactone. *Phytother Res* 17 (4): 420-1.

Yahara, S. *et al.* (1997). Taraxastane glycosides from *Eclipta alba. Phytochemistry* 44 (1): 131-135.

Zhang, M. *et al.* (1997). Isolation and identification of ecliptasaponin D from *Eclipta alba* (L.) *Hassk. Yao Xue Xue Bao* 32 (8): 633-4.

Zhang, M. and Chen, Y. (1996). Chemical constituents of *Eclipta alba* (L.) Hassk. *Zhongguo Zhong Yao Za Zhi* 21 (8): 480-1, 510.

BILVA (WOOD APPLE)

Botanical Name: *Aegle marmelos* Corr. ex Roxb.

Natural Order: Fabaceae

Active Ingredients: Mucilage, pectin, tannin, bitter principle, furanocoumarins: aegelinol, aurapten, marmin or marmelosin or imperatorin, marmelide, marmenol (Ali and Parvez 2004), psoralen, xanthotoxin, pyranocoumarin: luvangetin (Goel *et al* 1997), alkaloids: aegelin, aegelinin, haplopine, fagarine, skimmianine, minerals and protolimonoids: skimmiarepin A skimmiarepin C (Samarsekhra *et al* 2004).

Non-Clinical Studies: Anti-stress, anti-viral (Badam *et al* 2002), cardiotonic (Veerapan *et al* 1997), cardiac depressant (Hema and Lalitkumari 1988), uterine-relaxant (Hema and Lalitkumari 1988), anti-spermatogenic, wound healing, radio protective (Jagetia *et al* 2003), anti-ulcer (Goel *et al* 1997) and antimalarial (Shrinivas 2004).

BIBLIOGRAPHY

Ali, M.S. and Parvez, K.P. (2004). Marmenol: A 7-geranyloxy-coumarin from the leaves of *Aegle marmelos* Corr. *Natural Product Research* 18 (2): 141-146.

Badam, L. *et al.* (2002). In vitro antiviral activity of bael (*Aegle marmelos* Corr.) upon human coxsackieviruses B1-B6. *J Commun Dis* 34(2): 88-99.

Das, A.V.; Padayatii, P.S. and Paulose, C.S. (1996). Effect of leaf extract of *Aegle marmelos* (L.) Correa ex Roxb. on histological and ultra structural changes in tissues of streptozocin induced diabetic rats. *Indian J Exp Biol* 34: 341-5.

Goel, R.K. *et al.* (1997). Antiulcer activity of naturally occurring pyrano-coumarin and iso-coumarins and their effect on prostanoid synthesis using human colonic mucosa. *Indian J Exp Biol* 35: 1080-3.

Hema, C.G. and Lalitkumari, K. (1988). Screening of pharmacological actions of *Aegle marmelos*. *Indian Journal of Pharmacology* 20(2): 80-85.

Jagetia, G.C. *et al.* (2003). (Evaluation of the radio protective effect of *Aegle marmelos* (L.) Correa in cultured human peripheral blood lymphocytes exposed to different doses of gamena-radiation: a micronucleus study. *Mutagenesis* 18(4): 387-393.

Samarsekhra, J.K.R. *et al.* (2004), A new insecticidal protolimonoid from *Aegle marmelos*. *Natural Product Research* 18(2): 117-122.

Veerapan, A. *et al.* (1997) Cardio tonic effect of *Aegle marmelos* Corr. on amphibian heart in-situ preparation. *Poster:* 133. 6th Internet World Congress for Biomedical Sciences.

BISHOP'S WEED

Botanical Name: *Ammi visnaga* (L.) Lam.

Natural Order: Apiaceac

Active Constituents: Furochromones: khellin, khellol and its glycoside, visnagin. Pyranocoumarins: amidin, visnadin, flavonoids and volatile oil.

Non-Clinical Studies: Coronary vasodilator (Maurice and Walter (1950), inhibitor of oxalate induce nephrolithasis (Khan *et al* 2001), hypoglycemic (Joud *et al* 2002) and smooth muscle relaxant.

Clinical Studies

Angina Pectoris

Of 20 carefully studied patients with angina pectoris given oral visammin, 4 had a significant reduction in incidence of pain. Eleven stated that their pain was less severe and less easily precipitated while taking visammin. There was questionable improvement in the exercise tolerance test in 2 of 9 patients after oral visammin therapy. The daily oral dose ranged from 40 mg. to 240 mg. Undesirable side effects were encountered in 17 of the 20 patients. Parenteral administration of visammin was evaluated by means of the ballistocardiograph, the exercise tolerance test, and by tilting (Scott *et al* 1951).

Vitiligo

In order to evaluate the efficacy of topical khellin the vitiligo macules of one side only were painted in 41 patients with a 2% solution of khellin in acetone and propylene glycol (90 and 10%, respectively) and exposed to sunlight for a period of 4 months with 3 weekly applications and with exposure times up to 90 min. The macules of the other side were treated in 36 of the 41 patients with acetone and propylene glycol only and sun-exposed with the same schedule, while in the remaining 5 patients they were neither treated with khellin or placebo nor sun-exposed. No significant difference was evidenced between the khellin and placebo-treated sides: no excellent result (repigmentation more than 75% of the affected area) was found, and good results (repigmentation more than 50%) were found in 24.9% of khellin-plus sunlight-treated macules and in 22.3% of placeboplus sunlight-treated macules (Orecchia and Perfetti 1992).

Pharmacokinetics: After oral administration of 140 mg extract, khellin and visnadin were found in the plasma of six of the test subjects. No traces of visnagin were found; the maximum concentration of khellin was reached between 20 and 60 minutes at 29.4 and 276.5 ng/ml. Elimination occurred quickly; after 10 hours khellin was no longer detectable. For khellin the mean value was C_{max} value of 98.3 ng/ml. The LD_{50} of the extract administered orally is greater than 2000 mg/kg body weight for rats and mice.

After dosing rats with 10, 150 and 6000 mg extract/kg body weight over a period of four weeks, a minimal toxic level was reached at between 10 and 150 mg. After a 600 mg dose had been admini-stered a mild to medium grade centrolobular hypertrophy of the liver parenchyma with hepatocellular degeneration occurred.

Toxicity

Acute toxicity of aqueous extract: LD_{50} of intraperitoneal (i.p.) and oral administration was 3.6 and 10.1 g/kg in mice (Joud *et al* 2002).

LD50 of khellin: 80 mg/Kg.

Hepatotoxic and phototoxicity (due to presence of furochromones). After higher doses of khellin (100 mg daily administered orally), increased activity of the liver transaminases and of the gamma-glutamyl transferase have been observed.

BIBLIOGRAPHY

Duarte, J. *et al.* (1997). Effects of visnadine on rat vascular smooth muscle. In: *PM* 63 (3): 233-236.

Grienwald, R. and Stobernack, H.P. (1990). *Ammi visnaga*-Das Bischhofskraut. In: *ZPT* 11(2): 65.

Khan, Z.A. *et al.* (2001). Inhibition of oxalate nephrolithiasis with *Ammi visnaga* (Al-Khillah). *Int Urol Nephrol* 33 (4): 605-8.

Maurice, M. and Walter, S. (1950). Effect of Khellin on Coronary Artery Insufficiency as Evaluated by Electrocardiographic Tests. Circulation 2: 344.

Orecchia, G. and Perfetti, L. (1992). Photochemotherapy with topical khellin and sunlight in vitiligo. *Dermatology* 184 (2): 120-3.

Scott, R. *et al.* (1951). Studies on the effect of Oral and Parenteral Administration of Visammin (Khellin) in Patients with Angina Pectoris. *Circulation* 3: 80.

Joud, H. *et al.* (2002). Hypoglycemic effect of Aqueous extract of *Ammi visnaga* in Normal and Streptozotocin-Induced Diabetic Rats. *Journal of Herbal Pharmacotherapy*. 2 (4): 19-29.

BITTER APPLE

Botanical Name: *Citrullus colocynthis* (L.) Schard

Natural Order: Cucurbiataceae

Active Constituents: Glycoside: colocynthin.

Non-Clinical Studies: Drastic purgative, antifetility (Chaturvedi Mali and Ansari 2003).

BIBLIOGRAPHY

Anon: PDR for Herbal Medicines (2002). Medical Economics Company Inc, Montvale, NJ; 2000: 812-813.

Chaturvedi, M.; Mali, P.C. and Ansari, A.S. (2003). Induction of Reversible Antifertility with a Crude Ethanol Extract of *Citrullus colocynthis* Schrad Fruit in Male Rats *International Journal of Experimental and Clinical Pharmacology* 68(1): 38-48.

BLACK COHOSH

Botanical Name: *Cimicifuga racemosa* (L.) Nutt. Syn *Actaea racemosa* Linn. Syn *Cimicifuga cordifoliazz*

Natural Order: Ranuculaceae

Active Constituents: Triterpene glycosides or xylosides: actein, 27-deoxyactein, cimicifugoside M, cimiracemoside A, cimiracemoside B, cimiracemoside C, cimiracemoside D, cimiracemoside E, cimiracemoside F, cimiracemoside G, cimiracemoside H (Lindle 1968; Berger, Junior and Kopanski 1988; Anon: PDR for Herbal Medicines), phenylpropanoids (cimiracemate A, cimiracemate B), flavones (Struck Tegtmeier and Harnischfeger 1997), caffeic acid derivatives (Li *et al* 2003) and isoflavone (formononetin).

Non-Clinical Studies: Antiestrogenic (Zierau *et al* 2002; Liske *et al* 2002), antimalarial (Takahira *et al* 1998), estrogenic (Seidlova-Wuttke *et al* 2003), anticancer (Hemmi Kusano and Ishida 1980; Seidlova-Wuttke *et al* 2003) and agonist of serotonin receptor (Burdette *et al* 2003),

Clinical Studies

Hot flashes

A randomized trial reported efficacy of *Cimicifuga racemosa* for the treatment of hot flashes among women with a history of breast cancer (Jacobson *et al* 2001).

Toxicity: Hepatotoxic (Cohen *et al* 2004; Huntley 2004; Levitsky *et al* 2005).

BIBLIOGRAPHY

Berger, S.; Junior, P. and Kopanski, L. (1984). 27-Deoxyactein: A new polycyclic triterpenoid glycoside from *Actaea racemosa*. In: *Planta Med* 54: 579-580.

Beuscher, N. and Reichert, R. (1995). *Cimicifuga racemosa* L. - black cohosh. *Zeit Phytother* 16: 301-310.

Boblitz, N. *et al*. (2000). Benefit of a fixed drug combination containing St. John's wort and black cohosh for climacteric patients - results of a randomized clinical trial (poster presentation from 6th Annual Symposium on Complementary Health Care, Exeter, England, December 2-4, 1999). *Focus Alt Comp Ther* (FACT) 5 (1): 85-86.

Burdette, J.E. *et al*. (2003). Black cohosh acts as a mixed competitive ligand and partial agonist of the serotonin receptor. *J Agric Food Chem* 51 (19): 5661-5670.

Cohen, S.M. *et al*. (2004). Autoimmune hepatitis associated with the use of black cohosh: a case study. *Menopause* 11 (5): 575-577.

Hemmi, H., Kusano, G. and Ishida, N. (1980). Selective inhibition of nucleoside transport into mouse lymphoma L-5178Y cells by cimicfu-goside. *J Pharmacobiodyn* 3 (12): 636-642.

Jacobson, J.S. *et al*. (2001). Randomized trial of black cohosh for the treatment of hot flashes among women with a history of breast cancer. *J Clin Oncol*. 19 (10): 2739-2745.

Levitsky, J. *et al.* (2005). Fulminant liver failure associated with the use of black cohosh. *Dig Dis Sci* 50 (3): 538-9.

Li, W. *et al.* (2003). Identification of caffeic acid derivatives in *Actea racemosa* (*Cimicifuga racemosa*, black cohosh) by liquid chromatography/tandem mass spectrometry. *Rapid Commun Mass Spectrom*; 17 (9): 978-982.

Lindle, H. (1967). Contents of *Cimicifuga racemosa*. 2. On the structure of actein. *Arch Pharm Ber Dtsch Pharm Ges* 300 (10): 885-92.

Lindle, H. (1968). Contents of *Cimicifuga racemosa*. 4. Actein: Ring D and side chain. *Arch Pharm Ber Dtsch Pharm Ges* 301 (2): 120-38.

Liske, E. *et al.* (2002). Physiological investigation of a unique extract of black cohosh (*Cimicifugae racemosae* Rhizoma): a 6-month clinical study demonstrates no systemic estrogenic effect. *J Women's Health Gend Based Med* 11 (2): 163-174.

Seidlova-Wuttke, D. *et al.* (2003). Evidence for selective estrogen receptor modulator activity in a black cohosh (Cimicifuga racemosa) extract: comparison with estradiol-17beta. *Eur J Endocrinol* 149 (4): 351-362.

Struck, D.; Tegtmeier, M. and Harnischfeger, G. (1997). Flavones in extracts of *Cimicifuga racemosa*. *Planta Med* 63: 289-290.

Takahira, M. *et al.* (1998). Antimalarial activity and nucleoside transport inhibitory activity of the triterpenic constituents of *Cimicifuga* spp. *Biol Pharm Bull* 21 (8): 823-828.

Thomson, S. and Schimdt, M. (2002). Hepato toxicity from *Cimicifuga racemosa*? Recent Australian case report not sufficiently substantiated. *Med J Aus* 9 (3): 337-340.

Zierau, O. *et al.* (2002). Antiestrogenic activities of *Cimicifuga racemosa* extracts. *J Steroid Biochem Mol Biol* 80(1): 125-130.

BLUE COHOSH

Botanical Name: *Caulophyllum thalictroides* (L.) Michx

Natural Order: Berberidaceae

Active Constituents: Alkaloids: anagyrine, cytosine, magnoflorine (Ganzer *et al* 2003), saponins (Ganzer *et al* 2003) and glycoside of undetermined nature (Anon: PDR for Herbal Medicines).

Non-Clinical Studies: Ecbolic (Anon: PDR for Herbal Medicines), weak estrogenic and antispasmodic.

Toxicity: A 21-yr-old female developed tachycardia, diaphoresis, abdominal pain, vomiting and muscle weakness and fasciculations after using blue cohosh in an attempt to induce an abortion. These symptoms are consistent with nicotinic toxicity and probably resulted from methylcytisine known present in blue cohosh (Rao and Hoffmn 2002; Anon: PDR for Herbal Medicines).

BIBLIOGRAPHY

Anon. PDR for Herbal Medicines (2002). Medical Economics Company Inc, Montvale NJ 812-813.

Ganzer, M. *et al.* (2003). Determination of saponins and alkaloids in *Caulophyllum thalictroides* (blue cohosh) by high-performance liquid chromatography and evaporative light scattering detection. *Phytochem Anal* 14 (1): 1-7.

Rao, R.B. and Hoffmn, R.S. (2002). Nicotinic toxicity from tincture of blue cohosh (*Caulophyllum thalictroides*) used as an abortifacient. *Vet. Hum Toxicol* 44 (4): 221-2.

BRAHAMI

Botanical Name: *Bacopa monniera* (L.) Penn. syn *Herpestis monneira* (L.) H.B. and K.

Natural Order: Scophulariaceae

Active Ingredients: Steroid saponins: bacoside A, A1, A3, and B (Rastogi, Pal and Kulshreshtha Jain and Kulshreshtha 1992), alkaloids: herpestine and brahmine, dammarane-type saponins: bacopasaponins (Garai *et al* 1994; Garai *et al* 1996), betulic acid, phytosterols and flavonoid glycosides.

Non-Clinical Studies: Anticancer (Elangovan *et al* 1995), anti-inflammatory (Jain 1994), anxiolytic (Bhattacharya and Ghosal 1998), antioxidant (Tripathi *et al* 1996), muscle relaxant (Dar and Channa 1997), nootropic (Singh and Dhawan 1997), immuno-modulator (Dahanukar and Thatte 1997) and bronchodilator (Dar and Channa 1997).

Clinical Studies

Anxiety neurosis

The results of an open 4-week trial of bacopa in 35 patients with anxiety neurosis were published in 1990. The daily dose of dried bacopa herb was 12 g, as a syrup. Significant improvements in anxiety ($p<0.05$), concentration ($p<0.05$) and immediate memory span ($p<0.01$) were seen as a result of the treatment. Work-related mental fatigue, measured as total work output and errors committed per unit time, also improved significantly with bacopa treatment ($p<0.001$). Improvements were also seen in symptoms such as insomnia, headache, palpitation and irritability. No major side-effects were observed (Kidd 1999).

Mental functions

A. An Indian study investigated the effect of bacopa in school children aged 6-8 years. In a single-blind design, 40 children were given bacopa syrup equivalent to 1 g dried herb daily for 3 months. Immediate memory, perception and reaction/ performance times improved with bacopa treatment. No side-effects were observed (Kidd 1999).

B. An uncontrolled study of bacopa in patients suffering from epilepsy, mental retardation or mental illness suggested that a defatted extract of the dried plant was more therapeutically active than a decoction made from the fresh plant. There were some suggestion that the treatment could reduce the incidence of fits and improve other symptoms, but the results were inconclusive (Mukherjee and Dey 1996).

Epilepsy

In a subsequent open trial involving 13 epilepsy patients, administration of the defatted alcoholic bacopa extract for 3-16 months produced noticeable improvements in the frequency of epileptic fits in most patients. In 5 patients, fits were completely abolished, while in one patient the frequency of fits increased, (Kidd 1999).

Adverse drug reaction: Sedation (Anonymous, 1998).

BIBLIOGRAPHY

Anonymous (1998). *Indian Herbal Pharmacopoeia* Volume 1. Worli, Mumbai: Indian Drug Manufacturers Association

Bhattacharya, S.K. and Ghosal, S. (1998). Anxiolytic activity of a standardized extract of *Bacopa monniera* - an experimental study. *Phytomedicine* 5: 77-82.

Dar, A. and Channa S. (1997). Bronchodilatory and cardiovascular effects of an ethanol extract of *Bacopa monniera* in anaesthetized rats. *Phytomedicine* 4: 319-323.

Dahanukar, S.A and Thatte, U.M. (1997). Current status of Ayurveda in phytomedicine. *Phytomedicine* 4: 359-368.

Dar, A. and Channa, S. (1997). Relaxant effect of ethanol extract of *Bacopa monniera* on trachea, pulmonary artery and aorta from rabbit and guinea-pig. *Phytotherapy Research* 11: 323-325.

Elangovan, V. *et al.* (1995). *In vitro* studies on the anticancer activity of *Bacopa monnieri*. *Fitoterapia* 66: 211-215.

Garai, S. *et al.* (1996). Bacopasaponin D - a pseudojujubogenin glycoside from *Bacopa monniera*. *Phytochemistry* 43: 447-449.

Garai, S. *et al.* (1996). Dammarane-type triterpenoid saponins from *Bacopa monniera*. *Phytochemistry* 42: 815-820.

Jain, P. and Kulshreshtha, D.K. (1993). Bacoside A1, a minor saponin from *Bacopa monniera*. *Phytochemistr* 33: 449-451.

Jain, P. *et al.* (1994). Anti-inflammatory effects of an Ayurvedic preparation, *Brahmi Rasayan*, in rodents. *Indian Journal of Experimental Biology* 32: 633-636.

Kidd, P.M. (1999). A review of nutrients and botanicals in the integrative management of cognitive dysfunction. *Alternativ Medicine R*eview 4: 144-161.

Mukherjee, G.D. and Dey, C.D. (1966). Clinical trial on *Brahmi*. I. *Journal of Experimental Medical Sciences*; 10: 5-11.

Singh, H.K. and Dhawan, B.N. (1997). Neuropsychopharma-cological effects of the Ayurvedic nootropic *Bacopa monniera* Linn. (*Brahmi*). *Indian Journal of Pharmacology* 29: 359-S365.

Tripathi, Y.B. *et al.* (1996). *Bacopa monniera* Linn. as an antioxidant: mechanism of action. *Indian Journal of Experimental Biology* 34: 523-526.

BUCHU

Botanical Name: *Barosma beulina* (Berg.) Bartl. and Wendl. f. Syn *Agathosma betulina* Linn.

Natural Order: Rutaceae

Active Constituents: Essential oil: diosphenol, limonene, and menthone (Kaiser Lamparsky and Schudel 1975), bioflavonoids: rutin, barosmin or diosmin, hesperidin, sulphur.

Non-Clinical Studies: Antimicrobial (Kaiser, Lamparsky and Schudel 1975).

BIBLIOGRAPHY

Anon (2000). PDR for Herbal Medicines. Medical Economics Company Inc, Montvale, NJ 812-813.

Blommaert, K.L.J. and Bartel, E. (1976). Chemotaxonomic aspects of the buchu series *Agathosma betulina* Pillans and *A. crenulata* Pillans from local plantings. *Journal of South African Botany* 42 (2): 121-126.

Blumenthal, M., *et al.* (1998) (eds). The Complete German Commission E Monographs: Therapeutic Guide to Herbal Medicines. Austin: American Botanical Council and Boston: *Integrative Medicine Communications* 317.

Kaiser, R., Lamparsky, D. and Schudel, P. (1975). Analysis of buchu leaf oil. *Journal of Agricultural and Food Chemistry* 23 (5): 943-950.

Lis-Balchin, M. and Simpson, S.H. (2001). Buchu (*Agathosma betulina* and *A. crenulata*, Rutaceae) essential oils: their pharmacological action on guinea pig ileum and antimicrobial activity on micro-organisms. *Journal of Pharmacy and Pharmacology* 53 (4): 579-582.

Simpson, D. (1998), Buchu-South Africa's amazing herbal remedy. *Scott Med J* 43: 189–91 [review].

BUCKWHEAT

Botanical Name: *Fagopyrum esculentum* (Moench)

Natural Order: Chenopodiaceae.

Active Ingredients: Napthodianthrones: fagopyrin, protofagopyrin (Samel and de Witte 1995), flavonoids: rutin (Gressier *et al* 2000).

Non-Clinical Studies: Antioxidant (Gressier *et al* 2000) and peripheral vasodilator.

Clinical Studies

Chronic venous insufficiency

Sixty-seven male and female patients (22-74 years) with chronic venous insufficiency were randomly divided into two groups after a 2-week run-in period. They received either buckwheat herb tea (*Fagopyrum esculentum*) or a placebo tea for a period of 3 months. The main outcome measure was the lower leg volume determined by ultrasound. Subjective symptoms were assessed by a clinical symptom score system. The femoral vein diameters were measured by B-scan sonography.

In a subgroup of patients capillary permeability was determined by cutaneous fluorescence angiography. Although the mean partial leg volume did not change in the treatment group (from 2041 to 2073 ml), it increased in the placebo group by 110 ml (from 1972 to 2082 ml) according to intent to treat. The difference between the groups was significant. The subjective clinical symptoms were significantly reduced in both groups. The mean diameters of the femoral veins were reduced and capillary permeability was improved, but neither change was statistically significant. No drug-related adverse effects were observed. Chronic venous insufficiency is a very placebo-sensitive condition. The treatment with buckwheat herb tea is safe and could have a favourable influence on patients with chronic venous insufficiency such that further oedema development is prevented (Thme *et al* 1996).

BIBLIOGRAPHY

Gressier, B. *et al.* (2000). Phenolic compounds and antioxidant activities of buckwheat (*Fagopyrum esculentum* Moench) hulls and flour. *J Ethnopharmacol* 72 (1-2): 35-42.

Samel, D and de Witte p. (1995). Fagopyrins from *Fagopyrum esculentum* and their PTK inhibitory effect. In: *Planta Med* 61. (Absratcs of 43rd Ann Congr): 67.

Thme, N. *et al.* (1996). Leg oedema protection from a buckwheat herb tea in patients with chronic venous insufficiency: a single-centre, randomised, double-blind, placebo-controlled clinical trial. *Eu J Clin Pharmacol* 50 (6): 443-7.

BURDOCK

Botanical Name: *Arctium lappa* Linn.

Natural Order: Asteraceae

Active Constituents: Sesquiterpene lactone: arctiopicrin, bitter glucoside (lappatin or lappin), polyenes, sterols, carbohydrates: inulin, tannin: pholbaphene acids.

Non-Clinical Studies: Antibacterial, antifungal, anti-inflammatory (Lin *et al* 1996), antioxidant (Lin *et al* 1996; Lin *et al* 2002), antimutagenic (Morita, Kada and Namiki, 1984), antiplatelet (Iwakami *et al* 1992), hepaoprotective (Lin *et al* 2002), hypoglycemic Swanston-Flatt *et al* 1989) and ecbolic.

Adverse Drug Reations: Allergic dermatitis (Rodriguez *et al* 1995) and analphlaxis (Sasaki *et al* 2003).

BIBLIOGRAPHY

Iwakami, S. *et al.* (1992). Platelet activating factor (PAF) antagonists contained in medicinal plants: lignans and sesquiterpenes. *Chem Pharm Bull* (Tokyo) 40(5): 1196-1198.

Lin, C.C. *et al.* (1996). Anti-inflammatory and radical scavenge effects of *Arctium lappa*. *Am J Chin Med* 24: 127–37.

Lin, S.C. *et al.* (2002). Hepatoprotective effects of *Arctium lappa* Linn. on liver injuries induced by chronic ethanol consumption and potentiated by carbon tetrachloride. *J Biomed Sci* 9 (5): 401-9.

Morita, K.; Kada, T. and Namiki, M.A. (1984), desmutagenic factor isolated from burdock (*Arctium lappa* Linn). *Mutat Res* 129 (1): 25-31.

Rodriguez, P. *et al.* (1995). Allergic contact dermatitis due to burdock (*Arctium lappa*). *Contact Dermatitis* 33 (2): 134-135.

Sasaki, Y. *et al.* (2003). Anaphylaxis due to burdock. *Int J Dermatol* 42 (6): 472-473.

Swanston-Flatt, S.K. *et al.* (1989). Glycaemic effects of traditional European plant treatments for diabetes. Studies in normal and streptozotocin diabetic mice. *Diabetes Res* 10 (2): 69-73.

BURRAGE

Botanical Name: *Borago officinalis* Linn.

Natural Order: Boraginaceae

Active Constituents: Tannins, pyrrolizidine alkaloids: amabiline, polyunsaturated fatty acids: γ-linolenic acid and saponins.

Non-*Clinical Studies*: Anti-inflammatory (Kast (2001).

Clinical Studies

Atopic eczema

A. One hundred and sixty patients were randomized to take daily either 500 mg of borage oil-containing capsules or the bland lipid miglyol as a placebo over a 24-week period. Use of topical diflucortolone-21-valerate cream was allowed as rescue medication, with the amount used until response being defined as primary and clinical improvement as secondary efficacy criteria. Although several clinical symptoms improved compared with placebo, the overall response to borage oil did not reach statistical significance. Significant differences in favour of borage oil were, however, observed in a subgroup excluding patients who failed to show increased erythrocyte dihomo-gamma-linolenic acid levels and in whom adherence to inclusion criteria and the study protocol were questionable.

   GLA metabolites increased in borage oil-treated patients only, and serum IgE showed a trend to decrease on overall and subgroup analysis. No substance-related adverse effects were observed. This study shows no overall efficacy of GLA-containing borage oil in atopic eczema, with steroid use being the primary response parameter, although it suggests that a subgroup of patients may benefit from this well-tolerated treatment (Henz *et al* 1999).

B. 151 patients, of whom 11 failed to return for assessment, leaving an evaluable population of 140 (including 69 children). Adults received four capsules of borage oil twice daily (920 mg gamma linolenic acid), and children received two capsules twice daily, for 12 weeks. Change in total sign score at 12 weeks measured with the six area, six sign, atopic dermatitis (SASSAD) score (primary endpoint); symptom scores, assessed on visual analogue scales; topical corticosteroid requirement, assessed on a five point scale; global assessment of response by participants; adverse events and tolerability.

The mean SASSAD score fell from 30 to 27 in the borage oil group and from 28 to 23 in the placebo group. The difference between the mean improvements in the two groups was 1.4 (95% confidence interval -2.2 to 5.0) points in favour of placebo (P = 0.45). No significant differences occurred between treatment groups in the other assessments. Subset analysis of adults and children did not indicate any difference in response. The treatments were well tolerated. Gamma linolenic acid is not beneficial in atopic dermatitis (Takwale *et al* 2003).

Infantile seborrhoeic dermatitis

Details are missing (Tollesson and Frithz 1993).

Toxicity: Hepatotoxic.

BIBLIOGRAPHY

Anon (2002). PDR for Herbal Medicines. Medical Economics Company Inc, Montvale, NJ 812-813.

Henz, B.M. *et al.* (1999). Double-blind, multicentre analysis of the efficacy of borage oil in patients with atopic eczema. *Br J Dermatology* 140 (4): 685-8.

Kast, R.E. (2001). Borage oil reduction of rheumatoid arthritis activity may be mediated by increased cAMP that suppresses tumor necrosis factor-alpha. *International Immunopharmacol* 2197-99.1 (12): 2197-9.

Takwale A, *et al.* (2003). Efficacy and tolerability of borage oil in adults and children with atopic eczema: randomized, double blind, placebo controlled, parallel group trial. BMJ. 13; 327 (7428): 1385.

Tollesson, A, and Frithz, A. (1993). Borage oil, an effective new treatment for infantile seborrhoeic dermatitis. *Br J Derm* 12: 70.

CALIFORNIA POPPY

Botanical Names: *Eschscholtzia califronica* Linn.

Natural Order: Papaveraceae.

Active Ingredients: Alkaloids: escholzine (Fabre *et al* 2000), californidin (Fabre *et al* 2000), protopine (Fabre *et al* 2000), sangui-narine (Gertiq 1965; Jain *et al* 1997), 10-hydroxysanguinarine (Tanahashi and Zenk 1990), 12-hydroxychelirubine (Tanahashi and Zenk 1990), 10-hydroxychelerythrine (Tanahashi and Zenk 1990), 12-hydroxydihydrosanguinarine (Tanahashi and Zenk 1990), 12-hydroxydihydrochelirubine (Tanahashi and Zenk 1990), chelirubine, macarpine, dihydrosanguinarine (Tanahashi and Zenk 1990), dihydrochelirubine and dihydromacarpine (Tanahashiand Zenk 1990), escholinine (Necas *et al* 2002).

Non-Clinical Studies: Sedative.

BIBLIOGRAPHY

Fabre, N. *et al.* (2000). Direct characterization of isoquinoline alkaloids in a crude plant extract by ion-pair liquid chromatography-electrospray ionization tandem mass spectrometry: example of *Eschscholtzia californica*. *Journal of Chromatography*. 904 (1): 35-46.

Gertiq, H. (1965). Alkaloids of *Eschscholtzia californica* Cham. 3. Preparative separation of alkaloid fractions obtained from benzene root extract. *Acta Pol Pharm* 22 (3): 271-9.

Jain, L. *et al.* (1997). Alkaloids of *Eschscholtzia califronica Planta Med* 62(2): 188.

Necas, M. *et al.* (2002). (+)-Escholinine perchlorate. *Acta Cryst* E58, 1023-1024.

Tanahashi, T.K. and Zenk, M.H. (1990). New hydroxylated benzo[c]phenanthridine alkaloids from *Eschscholtzia californica* cell suspension cultures *Journal of Natural Product* 53(3): 579-86.

CANELLA

Botanical Name: *Canella alba* Linn.

Natural Order: Canellaceae

Active Constituents: Bitter constituent: canellin, resin.

Non-Clinical Studies: Antimicrobial (Anon: PDR for Herbal Medicines).

BIBLIOGRAPHY

Anon. (2002). *PDR for Herbal Medicines.* Medical Economics Company Inc, Montvale, NJ: 812-813.

CAT'S FOOT

Botanical Name: *Antennaria dioica* (L.) Gaertn. Syn. *Gnaphalium diocum* Linn.

Natural Order: Asteraceae

Active Constituents: Bitter glycosides, anthracene derivative and tannins.

Non-Clinical Studies: Antispasmodic and cholretic.

CHASTLE TREE

Botanical Name: *Vitex agnus castus* L.

Natural Order: Verbenaceae.

Active Ingredients: Irodoid glycosides, flavonoids: casticin (Hirbore *et al* 1997), glucosides: (Kuruuzum *et al* 2003).

Non-Clinical Studies: Cytotoxic (Hirbore *et al* 1997), prolactin inhibitor (Sliutz *et al* 1993; Jarry *et al* 1994).

Clinical Studies: Premenstural syndrome (Dittmar *et al* 1992; Lauritzen *et al* 1997; Loch Selle and Boblitz 2000; Schellenberg 2001), acne vulgaris (Amann 1967), mastlagia associated with menstrual cycle (Kubista Muller and Spona 1986; Wuttke *et al* 1997; Halaska *et al* 1999) and latent hyperprolactinemia (Milewicz *et al* 1993).

BIBLIOGRAPHY

Amann, W. (1967). Improvement of acne vulgaris following therapy with *Agnus castus* (Agnolyt). *Ther Ggw* 106: 124–126

Dittmar, F.W. *et al.* (1992). Premenstrual syndrome: Treatment with a phytopharmaceutical. *Therapiwoche Gynäkol* 1992; 5: 60-68.

Halaska, M. *et al.* (1999). Treatment of cyclical mastalgia with a solution containing a *Vitex agnus castus* extract: results of a placebo-controlled double-blind study. *Breast* 8: 175-181.

Hirbore, C. *et al.* (1997). Cytotoxic flavonoids from *Vitex agnus-castus*. *Phytochem* 46 (3): 521-524.

Jarry, H. *et al.* (1994). *In vitro* prolactin but not LH and FSH release is inhibited by compounds in extracts of *Agnus castus*: direct evidence for a dopaminergic principle by the dopamine receptor assay. *Exp Clin Endocrinol* 102: 448-454.

Kubista, E.; Muller, G. and Spona, J. (1986). Treatment of mastopathy associated with cyclic mastodynia: clinical results and hormone profiles [translated from German]. *Gynakol Rundsch* 26: 65-79.

Kuruuzum, A. *et al.* (2003). Glucosides from *Vitex agnus-castus*. *Phytochem* 63 (8): 959-964.

Lauritzen, C. *et al.* (1997). Treatment of premenstrual tension syndrome with *Vitex agnus-castus*. Controlled, double-blind study versus pyridoxine. *Phytomed* 4: 183–189.

Loch, E.G.; Selle, H, and Boblitz, N. (2000). Treatment of premenstrual syndrome with a phytopharmaceutical formulation containing *Vitex agnus castus*. *J Women Health Gender-Based Med* 9: 315–320.

Milewicz A, *et al.* (1993). *Vitex agnus-castus* extract in the treatment of luteal phase defects due to latent hyperprolactinemia. Results of a randomized placebo-controlled double-blind study [translated from German]. *Arzneimittelforschung* 43: 752–756.

Schellenberg, R. (2001). Treatment for the premenstrual syndrome with agnus castus fruit extract: prospective, randomized, placebo controlled study. *BMJ* 20: 134–137.

Sliutz G, *et al.* (1993). *Agnus castus* extracts inhibits prolactin secretion of rat pituitary cells. *Horm Metab Res* 25: 253–255.

Wuttke, W. *et al.* (1997). Treatment of cyclical mastalgia: results of a randomised, placebo-controlled, double-blind study [translated from German]. *Geburtsh Frauenheilk* 57: 569–574.

CHITRAKA (LEAD WORT)

Botanical Name: *Plumbago zeylanica* Linn. (White variety); *P. rosea* Linn. (Red variety).

Natural Order: Plumbaginaceae.

Active Ingredients: Napthoquinone: plumbagin (Vijayalaksmi, Rajan and Srinivasan 1987; Panichayupakaranant, Pharkphoom and Tewtrakul 2002; Satheesh and Seeni 2002), glucoside: hydroplumbagin, sterol: β-sitosterol and plumbagic acid.

Non-Clinical Studies: Antitumour (Fujji *et al* 1992; Singh *et al* 1997; Devi Solmon and Sharda 1999), antimicrobial (Paival *et al* 2003), anti-platelet (Shen *et al* 2003), central nervous system stimulant (Bopaiah and Prahdan 2001), hypolipidemic (Dwivedi 1997), antimutagenic (Durga Sridhar and Polasa 1992), radimodifying (Devi, Solmon and Sharda 1999).

Toxicity: The acute LD50 of plumbagin in normal mice: 9.4 mg/kg body weight.

BIBLIOGRAPHY

Bopaiah, C.P. and Prahdan, N. (2001). Central nervous system stimulatory action from the root extract of *Plumbago zeylanica* in rats. *Phytotherapy Res* 15 (2): 153-6.

Devi, U.; Solmon, F.E. and Sharda, A.C. (1999). Plumbagin, A Plant Napthoquinone with Antitumour and Radimodifying Properties. *Pharmaceutical Biology* 37 (3): 231-236.

Durga, R.; Sridhar, P. and Polasa, H. (1992). Antimutagenic activity of plumbagin in Ames *Salmonella typhimurium* test. *Indian Journal of Medical Research* 96: 143-5

Dwivedi, S. (1997). Effect of Plumbago zeylanica in hyperlipidemic rabbits and its modification by vitamin E. *Indian Journal of Pharmacology* 29 (2): 138.

Fujji, N. *et al.* (1992). Induction of topoisomerase II-mediated DNA cleavage by the plant naphthoquinones plumbagin and shikonin. *Antimicrob Agents Chemother* 36 (12): 2589–2594.

Singh, U.V. *et al.* (1997). Reduced toxicity and enhanced antitumor efficacy of plumbagin using poly (Lactic-co-glycolic) biodegradable injectable implant. *Indian Journal of Pharmacology* 29 (3): 168-172

Panichayupakaranant, Pharkphoom and Tewtrakul, Supinya (2002) Plumbagin production by root cultures of *Plumbago rosea*. *Electronic Journal of Biotechnology* 5 (3).

Paival, S.D. *et al.,* (2003). Antimicrobial Activity *In vitro* of Plumbagin Isolated from *Plumbago* Species. 959 *Mem Inst Oswaldo Cruz,* Rio de Janeiro 98 (7): 959-961.

Satheesh, K. and Seeni, S. (2002). Production of Plumbagin in Callus and Cell Suspension Cultures of *Plumbago indica* Linn. *Indian Journal of Biotechonology* 1 (3).

Shen, Z. *et al.,* (2003). Effects of plumbagin on platelet aggregation and platelet-neutrophil interactions. *Planta Med* 69 (7): 605-9

Sugie, S. *et al.,* (1998). Inhibitory effects of plumbagin and juglone on azoxymethane-induced intestinal carcinogenesis in rats *Cancer Lett* 127, 177-183.

Vijayalaksmi, J.; Rajan, S.S. and Srinivasan, R. (1987). Structure of plumbagin. *Acta Cryst* C43, 2375-2377

COMFREY

Botanical Name: *Symphytum officinale* (L.) Schard

Natural Order: Boraginaceae

Active Constituents: Pyrrolizidine alkaloids: acetyllycopsamine, echimidine, heliosupine intermedine, lasiocarpine, lycopsamine, symphytine, triterepene acids (rosmarinic acid), mucilage and allantoin.

Toxicity: Hepatotoxic (Ridker *et al* 1985; Weston *et al* 1987; Ridker and McDermott 1989; Ridker and McDermott 1989; Yeong *et al* 1990).

BIBLIOGRAPHY

Ridker, P.M. *et al.* (1985). Hepatic veno occlusive disease associated with the consumption of pyrrolizidine containing dietary supplements. *Gastroenterology* 88: 1050-1054.

Ridker, P.N. and McDermott, W.V. (1989). Comfrey herb tea and hepatic veno-occlusive disease. *Lancet* 1: 657-658.

Ridker, P.N. and McDermott, W.V. (1989). Hepatotoxicity due to comfrey herb tea. *Am J Med* 87: 701.

Roitman, J.N. (1981). Comfrey and liver damage. *Lancet.* 1:944.

Weston, C.F.M. *et al.* (1987). Veno-occlusive disease of the liver secondary to ingestion of comfrey. *Br Med J* 295:183.

Yeong, M.L. *et al.* (1990). Hepatic veno occlusive disease associated with comfrey ingestion. *J Gastroenterol Hepatol* 5: 211-4.

COMMON WITCH HAZEL

Botanical Name: *Hamamelis virginiana* L.

Natural Order: Hamamelidaceae.

Active Ingredients: Polymeric proanthocyanidins: hamameli-tannin and hamamelose (Haberland and Kolodziej 1993; Dauer *et al* 2003), flavonoids and volatile oil.

Non-Clinical Studies: Antiviral (Erdelmeier *et al* 1996), antipho-lgistic (Erdelmeier *et al* 1996), anti-inflammatory (Hartisch *et al* 1997; Hughes-Formella *et al* 1998).

BIBLIOGRAPHY

Dauer, A. *et al.* (2003). Polymeric proanthocyanidins from the bark of *Hamamelis virginiana. Planta Med* 69 (1): 89-91.

Erdelmeier, C.A.J. *et al.* (1996). Antiviral and antiphlogistic activities of *Hamamelis virginiana* bark. *Planta Med* 62 (3): 241–245.

Haberland, C. and Kolodziej, H. (1993). Novel galloylhamamelose from *Hamamelis virgiana.* In: *Planta Med* 59 (7) 8.

Hartisch, C. *et al.* (1997). Dual inhibitory activities of tannins from *Hamamelis virginiana* and related polyphenols on 5-lipoxygenase and lyso-PAF: acetyl-CoA acetyltransferase. *Planta Med* 63 (2): 106-10.

Hughes-Formella, B.J. *et al.* (1998). Anti-inflammatory effect of hamamelis lotion in a UVB erythema test. *Dermatology* 196 (3): 316–322.

CORNFLOWER

Botanical Name: *Centaurea cyanus* Linn.

Natural Order: Asteraceae

Active Constituents: Anthocyanidins, bitter glucosides of undetermined nature, flavonoids, indole alkaloids: moschamine, cis-moschamine, centcyamine and cis-centcyamine (Sarker *et al* 2001).

Non-Clinical Studies: Antibacterial (Anon: PDR for Herbal Medicines) and anti-inflammatory (Garbacki *et al* 1999).

BIBLIOGRAPHY

Garbacki, N. *et al.,* (1999). Anti-inflammatory and immunological effects of *Centaurea cyanus* flower-heads. *J. Ethnopharmacol* 68 (1-3): 235-41.

Sarker, S.D. *et al.,* (2001). Indole alkaloids from the seeds of *Centaurea cyanus* (Asteraceae). *Phytochemistry* 57 (8): 1273-6.

DAFFODIL

Botanical Name: *Narcissus pseudonarcissus* Linn.

Natural Order: Amayrllidacae.

Active Ingredients: Alkaloids: galanthine, galanthamine (Olin and Schneider 2001), haemanthamine, homolycorine, lycorine (Burrows and Tyrl 2001): Onfori *et al* 2003), masnine, narcidine (Tojo 1991), narcissine (Burrows and Tyrl 2001) and pluviin.

Non-Clinical Studies: Antiviral (Furusawa *et al* 1972), antileukemic (Furusawa *et al* 1972) and acetylcholinesterase inhibitor (Wake *et al* 2000).

Clinical Studies

Alzheimer's dementia

In a 6-month double-blind phase, 1289 patients were randomized to treatment with 24 or 32 mg/day of Galantamine vs. placebo, followed by a 6 month open-label phase, in which all patients received 24 mg/day. The group on Galantamine was benefited more than placebo in terms of Dementia scale (Olin and Schneider 2001).

Toxicity: Allergic dermatitis (Bruze Bjorkner and Hellstorm, 1996; Gude *et al* 1988).

BIBLIOGRAPHY

Bruze, M.; Bjorkner, B. and Hellstorm, A.C. (1996). Occupational dermatoses in nursery workers. *Am J Contact Dermat* 7(2):100-103.

Furusawa, E. *et al.* (1972). Effect of long-term administration of Narcissus alkaloid on Rauscher leukemia combinations with standard drugs. In: *Proc Soc Exp Biol Med* 140: 1034-1040.

Gude, M. *et al.* (1988). An investigation of the irritant and allergenic properties of daffodils (*Narcissus pseudonarcissus* L. Amaryllidaceae). A review of daffodil dermatitis. *Contact Dermatitis* 19 (1): 1-10.

Olin, J. and Schneider, L. (2001). Galantamine for Alzheimer's disease. Adult and Geriatric Treatment and Preventative Interventions Branch, National Institute of Mental Health. *Cochrane Database Syst Rev* 4: CD001747.

Onfori, S. *et al.* (2003). Effects of lycorine on growth and effects of L-galactonic acid-gamma-lactone on ascorbic acid biosynthesis in strains of *Cryptococcus laurentii* isolated from *Narcissus pseudonarcissus* roots and bulbs. *Antonie V Leeuwenkoek* 83(1): 57-61.

Suzuki, N. *et al.* (1974). Antiviral effect *in vitro* and rational drug combination in vivo. In *Proc Soc Exp Biol Med* 145: 771-777.

Tojo, E. (1991). (+)-Narcidine, a new alkaloid from *Narcissus pseudonarcissus*. In: *JNP* 54: 1387.

Wake, G. *et al.* (2000). CNS acetylcholine receptor actvity: *J Ethnopharmacol* 69: 105-14.

DARUHARIDRA (INDIAN BARBERRY)

Botanical Name: *Berberis aristata* DC.

Natural Order: Berberidiaceae

Active Ingredients: Alkaloids: berberine, berbamine, oxycanthine.

Non-Clinical Studies: Antiarrhythmic (Lau *et al* 2001), heparoprotective (Kulkarni *et al* 1972; Gilani and Janbaz 1992), cardio tonic (Gilani *et al* 1999), antiprotozoal (Gupe 1961; Subbaiah 1967; Soffar *et al* 2001), anticancer (Anis *et al* 2001), antibacterial (Amin, Subbaiah and and Abbasi 1969; Sack and Frohelich 1982), anti-trachoma (Sabir *et al* 1976), anti-chlamydial, antiheaprin (Sabir *et al* 1976), antidiarrhoeal (Bhide and Dutta, 1968), anti-inflammatory (Halder *et al* 1970; Kulkarni *et al* 1972), immunomodulator (Kulkarni *et al* 1972), antifungal (Nakamoto, Sadamori and Hamada 1990), antileukemic, immunosuppressant (Lou *et al* 1998) and uterine stimulant.

Clinical Studies

Diarrhea

1. A clinical study reported favorable response of berberine in acute watery diarrhoea (Khin-Maung *et al* 1985).
2. Series of 63 adult males with enterotoxigenic *E. coli* (ETEC) diarrhea of similar clinical characteristics were serially assigned to either an experimental group or a control group by use of a random number table. Thirty-three people were given a single dose of 400 mg berberine sulfate (BS) orally. The control group received no treatment. Both groups received IV rehydration therapy, and were observed for a period of 24 hours. During the total observation period, the experimental group had a 48% reduction in mean stool volumes compared with the controls ($p<0.05$). Also, when compared with the control group, significantly more of those receiving BS stopped having liquid diarrheal stools during the observation period (Rabbani *et al* 1987).

Ventricular tachyarrhythmias

The effects of berberine on 100 individuals with ventricular tachyarrhythmias observed with 24 to 48 hour ambulatory monitoring has also been reported. The results indicate 62% of patients had 50% or greater and 38% of patients had 90% or greater suppression of ventricular premature contractions (VPC). No severe side effects were observed; however, mild gastro-enterologic symptoms were reported by some patients (Huang 1990).

Trachoma

Fifty-one people attending an out-patient eye clinic having stage I or stage II trachoma lesions were recruited into the study. Subjects were divided into three groups: Group I received 0.2% berberine chloride eye drops, 2 drops per eye

3 times daily for 3 weeks; Group II received eye drops containing 0.2% berberine chloride plus 20% sulfacetamide according to the same schedule; Group III received eye drops containing only 20% sulfacetamide.

After 3 weeks, subjects in Group III had a slightly better clinical improvement as judged by conjunctival congestion, number of follicles and papillary reaction. The conjunctival scrapings of all subjects in Group III still tested positive for *C. trachomatis*, and these subjects were likely to have a relapse of symptoms. Subjects in Groups I and II showed a significant improvement in symptoms compared to their initial examination, and following the experimental period had only very mild symptoms remaining. All subjects in Groups I and II had conjunctival scrapings which were negative for *C. trachomatis*. Those in Groups I and II had no incidence of relapse up to one year later (Babbar *et al* 1982).

Toxicity: LD 50 of berberine sulfate in mice, i.v.: 25 mg/kg body weight. It is genotoxic (Pasqual *et al* 1993).

BIBLIOGRAPHY

Amin, A.H.; Subbaiah, T.V. and Abbasi, K.M. (1969). Berberine sulfate: antimicrobial activity, bioassay, and mode of action. *Can J Microbiol* 15: 1067-1076.

Anis, K.V. *et al.* (2001). Inhibition of chemical carcinogenesis by berberine in rats and mice. *J Physiol Pharmacol* 53 (5): 763-8.

Babbar, O.P. *et al.* (1982). Effect of berberine chloride eye drops on clinically positive trachoma patients. *Indian J Med Res* 76: 83-88.

Gilani, A.H. and Janbaz, K.H. (1992). Prevention of acetaminophen-induced liver damage by *Berberis aristata* leaves. *Biochem Soc Trans* 20 (4): 347S.

Gilani, A.H. *et al.* (1999). Possible mechanism of selective inotropic activity of the n-butanolic fraction from *Berberis aristata* fruit. *Gen Pharmacol* 33 (5): 407-14.

Gupe, S. (1961). Use of berberine in the treatment of giardiasis. *Am J of Diseases of Childhood*, 129: 866

Huang, W. (1990). Ventricular tachyarrhythmias treated with berberine *Chung Hua Hsin Hsueh Kuan Ping Tsa Chih*.18: 155-156,190.

Khin-Maung-U. *et al.* (1985). Clinical trial of berberine in acute watery diarrhoea. *Br Med J* 291:1601-1605.

Kulkarni, S.K. *et al.* (1972). Pharmacological investigations of berberine sulphate. *Japanese Journal of Pharmacology* 22 11-16,

Lau, C.W. *et al.* (2001). Cardiovascular actions of berberine. *Cardiovas Drug Rev* 19 (3): 234-44.

Lou, C.N. *et al.* (1998). Effect of berbamine on T-cell mediated immunity and the prevention of rejection on skin transplants in mice. *J Ethnopharmacol* 59 (3): 211-5.

Nakamoto, K.; Sadamori, S. and Hamada, T. (1990). Effects of crude drugs and berberine hydrochloride on the activities of fungi. *J Prosthet Dent* 64: 691-694.

Pasqual, M.S. *et al.* (1993). Genotoxicity of the isoquinoline alkaloid berberine in prokaryotic and eukaryotic organisms. *Mutat Res* 286: 243-252.

Rabbani G.H. *et al.* (1987). Randomized controlled trial of berberine sulfate therapy for diarrhea due to enterotoxigenic Escherichia coli and Vibrio cholerae. *J Inf Dis* 155: 979-984.

Sabir, M. *et al.* (1976). Experimental study of the antitrachoma action of berberine. *Indian J Med Res* 64: 1160-1167.

Sack, R.B. and Frohelich, J.L. (1982). Berberine inhibits intestinal secretory response of *Vibrio cholerae* and *Escherichia coli* enterotoxins. *Infect Immun* 35 (2): 471-5.

Soffar, S.A. *et al.* (2001). Evaluation of the effect of a plant alkaloid (berberine derived from *Berberis aristata* on *Trichomonas vaginalis in vitro*. *J. Egypt Soc Parasitol* 31 (3): 893-904.

Subbaiah, T.V. (1967). Effects of berberine sulphate on *Entamoeba histolytica*, *Natyuer* 215 527-28.

DEVDARU

Botanical Name: *Cedrus deodara* (Roxb.) Loud.

Natural Order: Pinaceae

Active Ingredients: Essential oil (contains sesquitrepene alcohol known as himachalol).

Non-Clinical Studies: Antiallergic (Gupta Tandon and Patnaik 1997).

BIBLIOGRAPHY

Gupta, P.P., Tandon, J.S. and Patnaik, G.K. (1997). Antiallergic activity of *Cedrus deodara*. *J Med Aromatic Plant Sci* 19: 1007-8.

DEVIL'S CLAW

Botanical Name: Harpagophytum procumbens DC.

Natural Order: Pedaliaceae.

Active Ingredients: Monoterpene glucosides: harpagide, harpagoside, procumbide (Pourrat *et al* 1985; Mestdagh and Torck 1995) triterpenoids, sterols and flavonoids.

Non-Clinical Studies: Analgesic (Mohmed and Ojewol 2004), anti-inflammatory (Mohmed and Ojewol 2004; Haung *et al* 2005), anti-arthritic (Andersen *et al* 2004), antidiabetic (Mohmed and Ojewol 2004), antiplasmodial (Clarkson *et al* 2003), negative chronotropic (Circosta *et al* 1984) and positive inotropic (Circosta 1984).

Toxicity: LD 50 of extract in mice: 13.5 mg/kg body weight. Harpagoside is reported to produce toxic symptoms when adminstered intravenous.

Clinical Studies

Myalgia

In a randomised, double-blind, placebo controlled study, *Harpagophytum procumbens* was investigated for efficicay in treating muscle pain. The study was conducted on patients with history of muscular pain in neck, shoulder and back. The test group received tablets of *Harpagophytum procumbens for twenty-eight days*. A total of 31 patients in the verum group and 32 in the placebo group were treated. *Harpagophytum procumbens* proved to be effective and no side effects were recorded (Gobel *et al* 2001).

Osteoarthritis

In a double-blind, randomized, multicentre clinical study, *Harpagophytum procumbens* was compared with diacerhein in the possible treatment of ostroarthritis. The study was conducted on one twenty two pateints and lasted for four months. At the end of the study, most of patients taking non-steroidal anti-inflammatory drugs have less dependency on them. Fewer aside effects were recored in *Harpagophytum procumbens* group (Chantre *et al* 2000).

Arthrosis of Hip Joint

In an uncontrolled multicentre drug surveillance study, *Harpagophytum procumbens* was investigated for efficacy in treating arthrosis of the hip or knee. The study lasted for tweleve weks and was conducted on seventy-five pateints. The authors reported favorable effect of *Harpagophytum procumbens* (Wegner and Lupke 2003).

BIBLIOGRAPHY

Andersen, M.L. *et al.* (2004). Evaluation of acute and chronic treatments with *Harpagophytum procumbens* on Freund's adjuvant-induced arthritis in rats. *J Ethnopharmacol* 91(2-3): 325-30.

Chantre, P. *et al.* (2000). Efficacy and tolerance of *Harpagophytum procumbens* versus diacerhein in treatment of osteoarthritis. *Phytomed* 7(3): 177-83.

Circosta, C. *et al.* (1984). A drug used in traditional medicine: *Harpagophytum procumbens* DC. II. Cardiovascular activity. *J Ethnopharmacol* 11(3): 259-74

Clarkson, C. *et al.* (2003). *In vitro* antiplasmodial activity of abietane and totarane diterpenes isolated from *Harpagophytum procumbens* (devil's claw). *Planta Med* 69 (8): 720-4.

Gobel, H. *et al.* (2001). Effects of *Harpagophytum procumbens* LI 174 (devil's claw) on sensory, motor und vascular muscle reagibility in the treatment of unspecific back pain. *Schmerz* 15 (1): 10-8.

Haung, T.H. *et al.* (2006). Harpagoside suppresses lipopolysa-ccharide-induced iNOS and COX-2 expression through inhibition of NF-kappaB activation. *J Ethnopharmacol* (Epub ahead of print).

Mestdagh, O. and Torck, M. (1995). Quality evaluation of Harpago-phyton capsules. Etude de la qualite de gelules d'Harpagophyton. *Ann Pharm Fr* 53 (3): 135-7

Mohmed, I.M. and Ojewol, J.A. (2004). Analgesic, anti-inflammatory and antidiabetic properties of *Harpagophytum procumbens* DC (Pedaliaceae) secondary root aqueous extract. *Phytother Res* 18 (12): 982-9.

Pourrat, H. *et al.,* (1985). Stability of iridoids of *Harpagophytum procumbens* DC. during the preparation of powders and atomized drugs. *Ann Pharm Fr* 43 (6): 601-6.

Wegner, T. and Lupke, N.P. (2003). Treatment of patients with arthrosis of hip or knee with an aqueous extract of devil's claw (*Harpagophytum procumbens* DC.). *Phytother Res* 17 (10): 1165-72.

EUROPEAN ANGELICA

Botanical Name: *Angelica archangelica* (Moench. Hoffm.).

Natural Order: Apiaceae

Active Constituents: Furanocoumarins: archangelicin, bergaptene, phellopterin, psoralen, xanthotoxin (Beyrich 1965; Glowniak *et al* 1996) hydroxycoumarins: scopoletin, essential oil.

Non-Clinical Studies: Antidepressant (Bergendorf *et al* 1997), antimutagenic (Salikhova and Poroshenko 1995), antioxidant (Yeh *et al* 2003), antiplatelet (Shimizu *et al* 1991), anti-ulcerogenic (Khayyal, *et al* 2001), hepatoprotective (Yeh *et al* 2003), viricidal (Sigurdsson Gudnadottir and Gudbjarnason).

Toxicity: LD 50 of Angelica seed oil in rats, orally > 5 mg/kg. Hepatotoxic and phototoxicity (due to presence of furanocoumarins).

BIBLIOGRAPHY

Bergendorf, O. *et al.* (1997). Furanocoumarins with affinity to brain benzodiazepine receptors *in vitro*. *Phytochemistry* 44: 1121-1124.

Beyrich, T. (1965). Isolation of phellopterin from the fruits of *Heracleum mantegazzianum* (Sommier et Levier) and *Angelica archangelica* L. 4. On furocoumarins. *Arch Pharm Ber Dtsch Pharm Ges* 298 (10): 672-6.

Glowniak, K. *et al.,* (1996). Localisation and seasonal changes of psoralen in Angelica fruits. *Planta Med* 62. Abstracts of the 44th Ann Congress of GA, 76.

Khayyal, M.T. *et al.* (2001). Antiulcerogenic effect of some gastro intestinally acting plant extracts and their combinations. *Arzneimittelforschung* 51: 545-553.

Salikhova, R.A. and Poroshenko, G.G. (1995). Antimutagenic properties *of Angelica archangelica* L. *Vestn Ross Akad Med Nauk.* 9(1): 58-61.

Shimizu, M. *et al.* (1991). Evaluation of *Angeliace radix* (Touki) by inhibitory effect on platelet aggregation. *Chem Pharm Bull* 39: 2046.

Yeh, M.L. *et al.* (2003). Hepatoprotective effect of *Angelica archan-gelica* in chronically ethanol-treated mice. *Pharmacology* 68(2):70-3.

EUROPEAN WILD GINGER

Botanical Name: *Asarum europaeum* L.

Natural Order: Aristolochiaceae

Active Constituents: Flavonoids, acids, essential oil: contains phenylpropanoids (Gracza 1980; Belova *et al* 1985) and camphor like principle.

Non-Clinical Studies: Emetic, antispasmodic, anesthetic, antibacterial and expectorant (Gracza 1981).

Clinical Studies

Acute and chronic bronchitis and bronchial asthma

In a double blind study comprising ninety patients, efficacy of *Asarum europaeum* was investigated in thirty patients of acute bronchitis, thirty patients of chronic bronchitis and thirty patients of bronchial asthma. Eighty per cent of the patients of acute bronchitis, fifty-eight per cent patients of chronic bronchitis and sixty-eight per cent case of bronchial asthma showed significant improvement.

Toxicity: LD50 of α-asarone for mice in enteral administration was 417.6 mg/kg, in intra-abdominal – 310 mg/kg (Belova *et al* 1985).

BIBLIOGRAPHY

Belova, L.F. *et al.* (1985). Asarone and its biological properties. *Farmakol Toksikol* 48 (6): 17-20.

Gracza, L. (1980). Phytobiological (phytopharmacological) studies on phenylpropane derivatives from *Asarum europaeum* L. 10th Communication: Active principles of *Asarum europaeum* (author's transl)]. *Arzneimittelforschung* 30 (5): 767-71.

Gracza, L. (1981). *In vitro* studies on the expectorant effect of phenylpropane derivatives from *Asarum europaeum* L. *Arzneimittelforschung* 30 (5): 767-771.

EVENING PRIMROSE OIL

Botanical Name: *Oenothera biennis* L.

Natural Order: Onagraceae

Active Constituents: Fatty oil: linoleic acid, γ-linoleic acid and oleic acid.

Non-Clinical Studies: Anti-rheumatic (Belch and Hill 2000), anti-platelet (De La Cruz *et al* 1997), antioxidant (De La Cruz*et al* 1999), gastro protective (al-Sabanah 1997),

Clinical Studies: Studies have revealed efficacy of evening prirmrose oil in atopic dermatitis (Berth-Jones 1993), hot flushes associated with PMS (Chenoy *et al* 1994), chronic mastitis (Blommers *et al* 2002), ulcerative colitis (Greenfield *et al* 1993), rheumatoid arthritis (Jantti *et al* 1989). A study reported inefficiency of evening primrose oil in chronic dermatitis of the hand. (Whitaker Cilliers and de Beer 1996).

BIBLIOGRAPHY

Berth-Jones, J. (1993). Placebo controlled trial of essential fatty acid supplementation in atopic dermatitis. *Lancet* 341: 1557-1560.

Blommers, J. *et al.* (2002). Evening primrose oil and fish oil for severe chronic mastalgia: a randomized, double-blind, controlled trial. *American Journal of Obstetrics and Gynecology* 187 (5): 1389-1394.

al-Sabanah, O.A. (1997). Effect of evening primrose oil on gastric ulceration and secretion induced by various ulcerogenic and necrotizing agents in rats. *Food Chem Toxicol* 35 (8): 769-775.

Belch, J.J. and Hill, A. (2000). Evening primrose oil and borage oil in rheumatologic conditions. *Am J Clin Nutr* 71: 352S-356S.

Chenoy, R. *et al.* (1994). Effect of oral gamma linolenic acid from evening primrose oil on menopausal flushing. *BMJ* 19 (308): 501-503.

De La Cruz, J.P. *et al.* (1997). Effect of evening primrose oil on platelet aggregation in rabbits fed an atherogenic diet. *Thromb Res* 87(1):141-149.

De La Cruz, J.P. *et al.* (1999). Antioxidant potential of evening primrose oil administration in hyperlipemic rabbits. *Life Sciences* 65(5): 543-555.

Greenfield, S.M. *et al.* (1993). A randomized controlled study of evening primrose oil and fish oil in ulcerative colitis. *Alimentary Pharmacology and Therapeutics* 7 (2): 159-166.

Jantti, J. *et al.* (1989). Evening primrose oil and olive oil in treatment of rheumatoid arthritis. *Clinical Rheumatology* 8 (2): 238-244.

Whitaker, D.K.; Cilliers, J. and de Beer. C. (1996). Evening primrose oil (Epogam) in the treatment of chronic hand dermatitis: disappointing therapeutic results. *Dermatology* 193 (2): 115-120.

EYEBRIGHT

Botanical Name: *Euphrasia officinalis* Linn.

Natural Order: Scrophulariaceae.

Active Ingredients: Irioid glycosides: acubin, adoxoside, catapol, euphroside, geniposide (Harkiss and Timmins 1973), flavonoids and tannins.

Non-Clinical Studies: Cytotoxic.

BIBLIOGRAPHY

Harkiss, K.J. and Timmins, P. (1973). Studies in the Scrophulariaceae. 8. Phytochemical investigation of *Euphrasia officinalis*. *Planta Med* 23 (4): 342-7.

FEVERFEW

Botanical Name: *Tanacetum parthenium* (L.) Schultz-Bip.

Natural Order: Asteraceae

Active Constituents: Diterpene lactone: parthenolide (Hausen and Osmundsen 1983; Heptinstall *et al* 1992; Nelson, Cobb and Shelton 2002).

Non-Clinical Studies: Anti-inflammatory, anti-leukemic, anti-platelet (Loesche, Groenewegen, Krause, Spangenberg and Heptinstall 1988), anticancer (Curry *et al* 2001; Won, Ong, Shi and Shen 2004).

Clinical Studies: Clinical studies have demonstrated feverfew as preventive and treatment of migraine (Waller and Ramsay 1985; Johnson *et al* 1985; Murphy, Heptinstall and Mitchell 1988; Murphy *et al* 1988; DeWeerdt, Bootsman and Hendricks 1996; Palevitch, Earon and Carasso 1997). A trial reported beneficial effect of feverfew in rheumatoid arthritis (Pattrick Heptinstall and Doherty 1989).

Adverse Drug Reactions: Contact dermatitis (Hausen and Osmundsen 1983).

BIBLIOGRAPHY

Johnson, E.S. *et al.* (1985). Efficacy of Feverfew as Prophylactic Treatment of Migraines. *British Medical Journal* 291: 569-573.

Murphy J.J., *et al.* (1988). Randomized double-blind placebo-controlled trial of Feverfew in Migraine Prevention. *Lancet* 2: 922-923.

Curry, E.A., III, Murry, D.J., Yoder, C., Fife, K., Armstrong, V., Nakshatri, H., O'Connell, M., and Sweeney, C.J. (2001) Phase I dose escalation trial of feverfew with standardized doses of parthenolide in patients with cancer. *Invest New Drugs*22 (3): 299-305.

DeWeerdt, C.J., Bootsman, H. and Hendricks, H. (1996). Herbal medicines in migraine prevention. Randomized double-blind placebo-controlled crossover trial of a feverfew preparation. *Phytomed* 3 (3): 225-230.

Hausen, B.M. and Osmundsen, P.E. (1983). Contact allergy to parthenolide in *Tanacetum parthenium* (L.) Schulz-Bip. (feverfew, Asteraceae) and cross-reactions to related sesquiterpene lactone containing Compositae species. *Acta DermVenereol* 63 (4): 308-314.

Loesche, W., Groenewegen, W.A., Krause, S., Spangenberg, P. and Heptinstall, S. (1988). Effects of an extract of feverfew (*Tanacetum parthenium*) on arachidonic acid metabolism in human blood platelets. *Biomed Biochim Acta* 47(10-11): S241-S243.

Murphy, J.J., Heptinstall, S. and Mitchell, J.R. (1988). Randomised double-blind placebo-controlled trial of feverfew in migraine prevention. *Lancet* 2 (8604):189-192.

Nelson, M.H., Cobb, S.E. and Shelton, J. (2002). Variations in parthenolide content and daily dose of feverfew products. *Am J Health Syst Pharm* 59 (16): 1527-1531.

Palevitch, D.; Earon, G. and Carasso, R. (1997). Feverfew (*Tanacetum parthenium*) as a prophylactic treatment for migraine: a double-blind placebo-controlled study. *Phytother Res* 11 (7): 508-511.

Pattrick, M., Heptinstall, S. and Doherty, M. (1989). Feverfew in rheumatoid arthritis: a double blind, placebo controlled study. *Ann Rheum Dis* 48 (7): 547-549.

Waller, P.C. and Ramsay, L.E. (1985). Efficacy of feverfew as prophylactic treatment of migraine. *Br Med J* 291 (6502): 1128.

Won, Y.K., Ong, C.N., Shi, X. and Shen, H.M. (2004). Chemopreventive activity of parthenolide against UVB-induced skin cancer and its mechanisms. *Carcinogenesis* 25 (8): 1449-1458.

FLOWERING DOGWOOD

Botanical Name: *Cornus florida* L.

Natural Order: Cornaceae.

Active Ingredients: Steroid saponins, iridoid glycoside: cornin, scyllitol.

Non-Clinical Studies: Antimalarial, antimitotic (Nisida and Murakami 1965) and antioxidant.

BIBLIOGRAPHY

Anon: (2002) PDR for Herbal Medicines. Medical Economics Company Inc, Montvale, NJ 12-813.

Nisida, I. and Murakami, T.H. (1965), Antimitotic action of cornin as a biologically active polypeptide. I. Biochemical properties of cornin. *Acta Med Okayama* 19 (1):1-9.

GINKGO

Botanical Name: *Ginkgo biloba* Linn.

Natural Order: Ginkgoaceae

Active Constituents: Flavonoids (isorhamnetin, kaempferol), bioflavonoids (5-methoxybilobetin, amentoflavone, bilobetin, ginkgetin, isoginkgetin, sciadopitysin), diterpenes: ginkgolide A, B, C, (Chung *et al* 1987; Corey *et al* 1988; Huh and Staba 1992; Joyeux *et al* 1995; Nakanishi, 1998; Lobstein-Guth *et al* 1998), trilactonic sesquiterpene: bilabolide (Huh and Staba 1992; Joyeux *et al* 1995), alkyl phenols: ginkgolic acid, hydro-ginkgolic acid, ginkgolinic acid, ginkgol, bilobol), proanthocyanidins: procyanidin, prodelphinidin, flavones (leutin, tricetin).

Non-Clinical Studies: Antilipoperoxidant, antinecrotic, antioxidant (Luqasi Horvahovich and Dworshack 1999) anticancer (Itokawa *et al* 1987; Kim 2001; Jane Chao, Chia Chou Chu 2004), anti-platelet (Chung *et al* 1987), neuro-protective (Krieglstein 1994), aphrodisiac (Cohen and Bartlik 1998), membrane stabilization, improvement of glucose utilization, anti-stress (Porsolt *et al* 1990), anti-ischemic (Mouren Caillard and Schwartz 1994), antidepressant (Subhan and Hindmarch 1984; Lingaerde, Foreland and Magnusson 1999), nitric oxide inhibitor, gastro protective, anti-inflammatory (Kwak *et al* 2002), anti-viral (Hayashi, Hayashi and Morita 1992), antagonist of cortical glycine and GABA $_A$ receptors (Lidiva, *et al* 2003).

Clinical Studies

Alzheimer's disease

A. In a randomized, double-blind, placebo-controlled trial two doses of *Ginkgo biloba* extract were tested in five hundred thirteen patients with dementia of the Alzheimer's type. The patients were instructed to consume *Ginkgo biloba* extract at daily doses of 120 mg or 240 mg or placebo for 26-weeks. The trial however reported ineffectiveness of *Ginkgo biloba* extract in dementia of the Alzheimer's type (Schneider *et al* 2005).

B. The U.S. trial enrolled more than 300 people with Alzheimer's disease or non-Alzheimer's dementia. Participants were given either 40 mg of *Ginkgo biloba* extract or placebo 3 times daily. The results showed significant (but not dramatic) improvements in the treated group.

C. In a randomized, placebo-controlled trial with *Ginkgo biloba* (180 mg per day) significant cognitive benefits were observed in healthy persons (Mix and Crews 2002).

D. A randomized, placebo-controlled trial with *Ginkgo biloba* failed to show efficacy in memory loss (Solomon *et al* 2002).

E. In a double-blind cross-over trial, one hour following a single oral 600 mg dose of *Ginkgo biloba* extract, short-term memory parameters were significantly improved compared with controls. According to authors, the trial indicated a specific activity on central cognitive processes and he proposes the use of *Ginkgo biloba* extract in cases and conditions where memory problems are a feature (Hindmarch 1988).

Erectile dysfunction

In a triple blind, placebo controlled study, efficacy of *Ginkgo biloba* was assessed in twenty-four patients with history of erectile dysfunction associated with antidepressants. Based on the observations statistically no significance difference was observed in the treatment and side-effects (Wheatley 2004).

Raynaunld's disease

A trial that compared dosages of 120 mg and 240 mg of ginkgo demonstrated a substantial therapeutic benefit on pain free walking distance with the higher dosage (Sikora *et al* 1998). Two placebo controlled trials, with a total of 190 patients, showed improved walking distance and decreased pain in patients with peripheral vascular disease (Bauer 1984; Peters, Kieser and Holscher 1998).

Schizophernia

In a study, fifty-four patients with chronic refractory schizophrenia, twenty-seven were treated with haloperidol plus

Ginkgo biloba, and the rest received haloperidol plus placebo. Based on the observations it was concluded that *Ginkgo biloba* may enhance the efficacy of haloperidol on schizophrenia, especially on positive symptoms (Zhou *et al* 1999).

Tinnitus

A. A recent study of 1,121 subjects conducted using questionnaires and telephone interviews, without the use of standard audiometric testing as an outcome measure, failed to show a benefit of ginkgo in the treatment of tinnitus (Drew and Davies 2001).

B. Another randomized, placebo-controlled trial of one hundred three patients showed 50 percent of patients with new-onset tinnitus had improvement or disappearance of symptoms in 70 days compared with 119 days to improvement in those receiving placebo (Drew and Davies 2001).

Ulcerative colitis

An open trial of cedemin, a *Ginkgo biloba* extract with PAF-antagonistic effects for ulcerative colitis has been reported but details are missing (Sandberg-Gertzén 1993).

Bronchial asthma

A study reported effect of concentrated oral liquor of *Ginkgo biloba* in the treatment of bronchial asthma but details are missing (Li *et al* 1997).

Hypoxia

A double blind study reported efficacy of *Ginkgo biloba* in protecting against hypoxia (Schaffler and Reeh 1985).

Macular degeneration

A study reported favorable effect of *Ginkgo biloba* in senile macular degeneration (Lebuuison, Leroy and Rigal 1986).

Common cold

A study reported favorable effect of *Ginkgo biloba in* mountain sickness and vascular reactivity to cold exposure (Roncin, Schwartz and D'Arbigny 1996).

Depression

In an open study, polysomnographic effect of *Ginkgo biloba* as an adjuvant therapy with tricyclic antidepressant, trimipramine was investigated in patients suffering from severe depression. The authors reported favorable response of Ginkgo *biloba* as an adjuvant to trimpramine in treating depression (Hemmeter *et al* 2001).

Pharmacodynamics and Kinetics: Pharmacotherapeutic evaluation of *Ginkgo biloba* extract at 120 mg BID showed no significant effect on BP, HR, or EKG parameters over 7 days in a randomized, placebo-controlled, crossover study of healthy volunteers (n=15). After oral administration, the bioavailability of ginkgolide A was >80% (half-life 4 hours); 70% excreted unchanged in urine. The bioavailability of ginkgolide B was >80% (half-life 6 hours); 50% excreted unchanged in urine. Bilobalide was 70% bioavailable with a half-life of 3 hours after 120 mg GBE ingestion with 30% excretion in urine. *Ginkgo biloba* extract did not have any effect on digoxin pharmacokinetics in 8 healthy volunteers (Kressman 2002).

Contra indications: Hypersensitivity.

Drug Interactions: Antiplalet drugs like aspirin and clopidogrel. The chief constituents of *Ginkgo biloba* including terpene trilactones and flavonol glycosides do not significantly inhibit these human CYPs *in-vitro*. However, amentoflavone and several other non-glycosidic constituents are significant *in-vitro* inhibitors of CYP (Moltke *et al* 2004).

Adverse Drug Reactions: Gastric upset (Stalleicken and Ihm 1989), hemorrhage (Skogh 1998), nephropathy (Lin and HO 1994), diarrhea, allergic skin rash or dermatitis (Sowers *et al* 1965), rarely spasms, cramps and atonia and adynamia (PDR for Herbal Remedies), hyphema (Rosenblatt and Mindel 1997).

Special Precautitions: Bleeding diathesis (Rowin and Lewis 1996; Vale 1998).

Toxicity: Allergenic, embryo toxic (Baron-Ruppert and Luepke 2001), immunotoxic (Koch, Jaggy and Chatterjee 2000). LD_{50} 1.8 mg/egg; 33 ppm for fraction containing ginkgolic acids

16% and biflavones 6.7%; LD_{50} 3.5 mg/egg; 64 ppm for fraction containing 58% ginkgolic acids but less than 0.02% biflavones; LD_{50} 250 mg/egg; 4540 ppm for fraction contai-ning 1% ginkgolic acids but less than 16% biflavones (Baron-Ruppert and Luepke 2001). The LD50 doses for *Ginkgo biloba* preparations as inhibitors of lipid peroxidation: 1.20 ± 0.06 mg/ml (Boveris and, Puntarulo 1998). LD 50 for the extract of *Ginkgo biloba* is 15.3 g/Kg. The extract of *Ginkgo biloba* is free from mutagenicity. According to studies conducted at National Toxicological Program, acute toxicity data is: oral LD_{50} in mice of 7.73 g/kg (1.9g/kg of flavone glycosides, 464 mg/kg of terpene glycosides), I.V. LD_{50} 1.1g/kg. According to American Botanical Council, LD50 of the *Ginkgo biloba* extract in the mouse was 7725 mg/kg body weight after oral application and 1100 mg/kg body weight after intravenous application (Foster 1991). LD 50 for acetone-water extract in mice: 7.7 g/Kg oral; 1.1 g/Kg i.p., in rats 2.1 g/Kg i.p.; 1.1 g/Kg i.v. (European Agency for the evaluation of medicinal products).

BIBLIOGRAPHY

Moltke, L.L.V. *et al.* (2004). Inhibition of human cytochrome P450 by components of *Ginkgo biloba*. *Journal of Pharmacy and Pharmacology*. 56(8), 1039.

Baron-Ruppert, G. and Luepke, N.P. (2001). Evidence for toxic effects of alkylphenols from *Ginkgo biloba* in the hen's egg test (HET). *Phytomedicine*; 8(2):133-8.

Bauer, U. (1984). 6-Month double-blind randomized clinical trial of *Ginkgo biloba* extract versus placebo in two parallel groups in patients suffering from peripheral arterial insufficiency. *Arzneimittelforschung*; 34:716-20.

Boveris, A.D. and Puntarulo, S. (1998). Free radical scavenging actions of natural antioxidants. *Nut Res* 18:1545-1557

Chung, K.F. *et al.* (1987). Effect of a ginkgolide mixture (BN 52063) in antagonizing skin and platelet responses to platelet activating factor in man. *Lancet* 31: 248-251.

Cohen, A.J. and Bartlik, B. (1998). *Ginkgo biloba* for antidepressant-induced sexual dysfunction. *Journal of Sex and Marital Therapy* 24: 139-143.

Cory, E.J. *et al.* (1988). Total Synthesis of Ginkgolide B. *J. Am. Chem. Soc.* 110: 649-651.

Drew, S. and Davies, E. (2001). Effectiveness of *Ginkgo biloba* in treating tinnitus: double blind, placebo controlled trial. *BMJ* 322 (7278): 73.

Hemmeter, U. *et al.* (2001). Polysomnographic effect of *Ginkgo biloba* as an adjuvant therapy in patients with major depression medicated with trimipramine. *Pharmacopsychiatry* 34 (2): 50-9.

Hindmarch, I. (1988). Activity of *Ginkgo biloba* Extract on Short-term Memory. E.W. Funfgeld (ed.) 1988. *Op cit* 321-326.

Hofferberth, B. (1994). The efficacy of EGb 761 in patients with senile dementia of the Alzheimer type, a double-blind, placebo-controlled study on different levels of investigation. *Human Psychopharmacology* 9: 215-222.

Huh, H. and Staba, E.J. (1992). The botany and chemistry of *Ginkgo biloba* L. *Journal of Herbs, Spices and Medicinal Plants* 1: 92-124.

Foster, S. (1991). *Ginkgo biloba. Botanical Booklet Series,* No. 304. Austin, TX: American Botanical Council.

Itokawa, H., *et al.* (1987). Antitumor principles from *Ginkgo biloba* L. *Chemical and Pharmaceutical Bulletin* 35: 3016-3020.

Jane, C.J., Chao, Chia and Chou Chu (2004). Effects of *Ginkgo biloba* extract on cell proliferation and cytotoxicity in human hepato-cellular carcinoma cells. *World Journal of Gastroenterology* 10 (1): 37-41

Jaqqy and Koch, E. (1997). Chemistry and biology of alkylphenols from Ginkgo *biloba* L. *Pharmazie* 52 (10): 735-8.

Joyeux, M. *et al.* (1995). Comparative Antilipoperoxidant, antinecrotic and scavenging properties of terpenes and biflavones from Ginkgo and some flavonoids. *Planta Med* 61 (2): 126-129.

Hayashi, K.; T. Hayashi and N. Morita (1992). Mechanism of action of the antiherpesvirus biflavone ginkgetin. *Antimicrobial Agents and Chemotherapy* 36(9): 1890-1893.

Kressmann, S. *et al.* (2002). Influence of pharmaceutical quality on the bioavailability of active components from Ginkgo biloba preparations. *J. Pharm Pharmacol.* 54(11): 1507-1514.

Kanowski, S. *et al.* (1995). Proof of efficacy of the *Ginkgo biloba* special extract Egb 761 in outpatients suffering from primary degenerative dementia of the Alzheimer type and multi-infract dementia. *Pharmacopsychiatry* 4: 149-158.

Kim, Y.S. *et al.* (1998). Antiplatelet and antithrombotic effects of a combination of ticlopidine and *Ginkgo biloba* ext (EGb 761) *Thromb Res* 91 (1): 33-8.

Kim, S.J. (2001). Effect of biflavones of *Ginkgo biloba* against UVB-induced cytotoxicity *in vitro*. *J Dermatol* 28 (4): 193-9.

Kleijnen, J. and Knipschild P. (1992). *Ginkgo biloba* for cerebral insufficiency. *Br. J Clin Pharmac* 35: 352-358.

Koch, E.; Jaggy, H. and Chatterjee, S.S. (2000). Evidence for immunotoxic effects of crude *Ginkgo biloba* L. leaf extracts using the popliteal lymph node assay in the mouse. *Int. J Immunopharmacol* 22 (3): 229-36.

Krieglstein, J. (1994). Neuroprotective properties of *Ginkgo biloba* constituents. In: *ZPT* 15 (2): 92-96.

Kwak, W.J. *et al.* (2002). Effects of Ginkgetin from *Ginkgo biloba* leaves on cyclooxygenases and in vivo skin inflammation. *Planta Medica* 68 (4): 316-21.

Lebuuison, D.A.; Leroy, L. and Rigal, G. (1986). Treatment of senile macular degeneration with Ginkgo biloba extract. A Preliminary double-blind drug vs. placebo study. *Presse Med* 15 (31): 1556-8.

Li, M.H. *et al.* (1997). Effects of ginkgo leave concentrated oral liquor in treating asthma, *Chung Kuo Chung Hsi I Chieh Ho Tsa Chih* 17(4): 216-8

Lidiva, I. *et al.* (2003). Terpene trilactones from *Ginkgo biloba* are antagonists of cortical glycine and GABAA receptors. *J. Biol. Chem*; 278(49), 49279-49285.

Lin, J.L. and Ho, Y.S. (1994). Flavonoid-induced acute nephropathy. *Am J Kidney Dis* 23 (3): 433-40.

Lingaerde, O., Foreland, A.R. and Magnusson, A. (1999). Can winter depression be prevented by *Ginkgo biloba* extract? A placebo-controlled trial. *Acta Psychiatrica Scandinavica* 100: 62-66.

Lobstein-Guth, A. *et al.*, (1998). Isolation of amentoflavone from *Ginkgo biloba*. *Planta Med* 54 (6): 555-6.

Luqasi, A.; Horvahovich, P. and Dworshack, E. (1999). Additional information to the *In vitro* antioxidant activity of *Ginkgo biloba* L. *Phytother Res* 13 (2): 160-2.

Marcilhac, A. *et al.* (1998). Effect of chronic administration of *Ginkgo biloba* extract or ginkgolide on the hypothalamic-pituitary-adrenal axis of the rat. *Life Sciences* 62: 2329-2340.

Meyer, B. (1986). Multicenter randomized double-blind drug vs. placebo study of the treatment of tinnitus with *Ginkgo biloba* extract [in French]. *Presse Med* 15:1562-4.

Mix, J.A. and Crews, W.D. Jr. (2002). A double-blind, placebo-controlled, randomized trial of *Ginkgo biloba* extract EGb 761 in a sample of cognitively intact older adults: neuropsychological findings. *Hum Psychopharmacol* 17: 267-77.

Mouren, X., Caillard, P. and Schwartz, F. (1994). Study of the anti-ischemic action of EGb 761 in the treatment of peripheral arterial occlusive disease by TcPo2 determination. *Angiology* 45: 413-417.

Nakanishi, K. (1988). Ginkgolides - Isolation and Structural Studies Carried out in the Mid-1960s. *P. Braquet* (ed.) Op cit 27-36.

Oken, B.S., Storzbach, D.M. and Kaye, J.A. (1998). The efficacy of *Ginkgo biloba* on cognitive function in Alzheimer's disease. *Archives of Neurology* 55: 1409-1415.

Peters, H., Kieser, M. and Hölscher, U. (1998). Demonstration of the efficacy of *Ginkgo biloba* special extract EGb 761® on intermittent claudication – a placebo-controlled, double-blind multicenter trial. *VASA* 27: 106-110.

Peters, H.; Kieser, M. and Holscher, U. (1998). Demonstration of the efficacy of *Ginkgo biloba* special extract EGb 761 on intermittent claudication — a placebo-controlled, double-blind multicenter trial. *Vasa*; 27:106-10.

Porsolt, R.D. *et al.*, (1990). Effects of an extract of *Ginkgo biloba* (EGB 761) on "Learned Helplessness" and other models of stress in rodents. *Pharmacology, Biochemisty and Behavior* 36: 963.

Roncin, J.P.; Schwartz, F. and D'Arbigny, P. (1996). EGb 761 in control of acute mountain sickness and vascular reactivity to cold exposure. *Aviat Space Environ Med* 67:445-52.

Rosenblatt, M. and Mindel, J. (1997). Spontaneous hyphema associated with ingestion of *Ginkgo biloba* extract. *N Engl J Med* 336:1108.

Rowin, J. and Lewis, S.L. (1996). Spontaneous bilateral sub-dural hematomas associated with chronic *Ginkgo biloba* ingestion. *Neurology* 46: 1775-1776.

Sandberg-Gertzén, H. (1993). An open trial of cedemin, a *Ginkgo biloba* extract with PAF-antagonistic effects for ulcerative colitis. *American Journal of Gastroenterology* 88: 615-616 (letter).

Schaffler, V.K. and P.W. Reeh. (1985). Double-blind Study of the Hypoxia-protective. Effect of a Standardized *Ginkgo biloba* Preparation after Repeated Administration in Healthy Volunteers. *Arzneim-Forsch* 35: 1283-1286.

Schneider, L.S. *et al.* (2005). A randomized, double-blind, placebo-controlled trial of two doses of *Ginkgo biloba* extract in dementia of the Alzheimer's type. *Curr. Alzheimer Res.* 2 (5): 541-51.

Schweizer, J. and Hautmann, C. (1999). Comparison of two dosages of *Ginkgo biloba* extract EGb 761 in patients with peripheral arterial occlusive disease Fontaine's stage IIb. A randomised, double-blind, multicentric clinical trial. *Arzneimittelforschung* 49: 900-4.

Sikora, R. *et al.* (1998). Randomized placebo-controlled study on the effects of oral treatment with *Ginkgo biloba* extract in patients with erectile dysfunction. *The Journal of Urology* 159 (5): 240 (abstract).

Skogh, M. (1998). Extracts of *Ginkgo biloba* and bleeding or haemorrhage. *The Lancet* 352: 1145-1146.

Solomon PR *et al.* (2002). Ginkgo for memory enhancement: a randomized controlled trial. JAMA 288: 835-40.

Sowers, S. *et al.* (1965). Ginkgo tree dermatitis. *In: Arch Dermatol* 81: 452-456.

Stalleicken, D. and Ihm, P. (1989). Observation of the Course of Cognitive Deficits. Results of a Multicenter Study Involving Psychological Test Operations. *Neurologie Psychiatrie* Special Issue 1: 64-69.

Subhan, Z. and Hindmarch, I. (1984). The psychopharmacological effects of *Ginkgo biloba* extract in normal healthy volunteers. *International Journal of Clinical and Pharmacological Research* 4: 89-93.

Vale, S. (1998). Subarachnoid hemorrhage associated with *Ginkgo biloba. Lancet* 352: 36.

Vesper, J. and Hansgen, R. (1994). Efficacy of *Ginkgo biloba* in 90 out patients with cerebral insufficiency caused by old age. *Phytomedicine* 1: 9-16.

Vorberg, G. (1985). *Ginkgo biloba* extract (GBE): a long-term study of chronic cerebral insufficiency in geriatric patients. *Clinical Trials Journal* 22: 149-157.

Wheatley, D. (2004). Triple-blind, placebo-controlled trial of *Ginkgo biloba* in sexual dysfunction due to antidepressant drugs. *Hum Psychopharmacol* 19 (8): 545-8.

Winther, K. *et al.* (1998). Effects of *Ginkgo biloba* extract on cognitive function and blood pressure in elderly subjects. *Current Therapeutic Research* 59: 881-888.

Zhou, D. *et al.* (1999). The effects of classic antipsychotic haloperidol plus the extract of *Ginkgo biloba* on superoxide dismutase in patients with chronic refractory schizophrenia. *Chin Med J* (Engl.) 112 (12): 1093-6.

GLOBE ARTICHOKE

Botanical Name: *Cynara scolymus* Linn.

Natural Order: Asteraceae.

Active Ingredients: Sesquiterpene lactones: cynarin, cynaropicrin, dehydrocynaropicrin, cynanratriol, grossheimin, flavonoids: rutin (Wang *et al* 2003) and inulin.

Non-Clinical Studies: Antioxidant (Wang *et al* 2003). cholretic, hypolipidemic (Luppatelli *et al* 2004), antispasmodic (Emendorfer *et al* 2005), antifungal (Zhu*et al* 2005) and hepatoprotective (Maros *et al* 1966).

Clinical Studies

Spastic Colon

Two hundred and eight (208) adults were identified post hoc as suffering with IBS. IBS incidence, self-reported usual bowel pattern, and the Nepean Dyspepsia Index (NDI) were compared before and after a 2-month intervention period. There was a significant fall in IBS incidence of 26.4% ($p < 0.001$) after treatment. A significant shift in self-reported usual bowel pattern away from "alternating constipation/diarrhea" toward "normal" ($p < 0.001$) was observed. NDI total symptom score significantly decreased by 41% ($p < 0.001$) after treatment (Bundy *et al* 2004).

Alchol induced hangovers

A study recruited healthy adult volunteers between 18 and 65 years of age to participate in a randomized double-blind crossover trial. Participants received either 3 capsules of commercially available standardized artichoke extract or indistinguishable, inert placebo capsules immediately before and after alcohol exposure. After a 1-week washout period the volunteers received the opposite treatment. Participants predefined the type and amount of alcoholic beverage that would give them a hangover and ate the same meal before commencing alcohol consumption on the 2 study days.

The primary outcome measure was the difference in hangover severity scores between the artichoke extract and placebo interventions. Secondary outcome measures were differences between the interventions in scores using a mood profile questionnaire and cognitive performance tests administered 1 hour before and 10 hours after alcohol exposure. Fifteen volunteers participated in the study.

The mean number (and standard deviation) of alcohol units (each unit being 7.9 g, or 10 mL, of ethanol) consumed during treatment with artichoke extract and placebo was 10.7 (3.1) and 10.5 (2.4) respectively, equivalent to 1.2 (0.3) and 1.2 (0.2) g of alcohol per kilogram body weight. The volume of nonalcoholic drink consumed and the duration of sleep were similar during the artichoke extract and placebo interventions. None of the outcome measures differed significantly between interventions. Adverse events were rare and were mild and transient. Our results suggest that artichoke extract is not effective in preventing the signs and symptoms of alcohol-induced hangover (Pittler *et al* 2004).

Functional dyspepsia

In a double-blind, randomized controlled trial, 247 patients with functional dyspepsia were recruited and treated with either a commercial ALE preparation (2 x 320 mg plant extract t.d.s.) or a placebo. The primary efficacy variable was the sum score of the patient's weekly rating of the overall change in dyspeptic symptoms (four-point scale). Secondary variables were the scores of each dyspeptic symptom and the quality of life (QOL) as assessed by the Nepean Dyspepsia Index (NDI). Two hundred and forty-seven patients were enrolled, and data from 244 patients (129 active treatment, 115 placebo) were suitable for inclusion in the statistical analysis (intention-to-treat).

The overall symptom improvement over the 6 weeks of treatment was significantly greater with ALE than with the placebo (8.3 ± 4.6, vs. 6.7 ± 4.8, $P < 0.01$). Similarly, patients treated with ALE showed significantly greater improvement in the global quality-of-life scores (NDI) compared with the placebo-treated patients (- 41.1 +/-??? 47.6 vs. - 24.8 +/- 35.6, $P < 0.01$). The ALE preparation tested was significantly better than the placebo in alleviating symptoms and improving the disease-specific quality of life in patients with functional dyspepsia (Holttman *et al* 2003).

BIBLIOGRAPHY

Wang, M. *et al.* (2003). Analysis of antioxidative phenolic compounds in artichoke (*Cynara scolymus* L.). *J Agr food Chem* 51 (3): 601-8.

Maros, T. *et al.* (1966). Effects of *Cynara scolymus* extracts on the regeneration of rat liver. *Arzneimittelforschung* 16 (2): 127-9

Holttman, G. *et al.* (2003). Efficacy of artichoke leaf extract in the treatment of patients with functional dyspepsia: a six-week placebo. *Aliment Pharmacol Ther* 18 (11-12): 109-115.

Pittler, M.H. *et al.* (2004). Effectiveness of artichoke extract in preventing alcohol-induced hangovers: a randomized controlled trial. *CMAJ* 170 (8): 1207-8.

Emendorfer, F. *et al.* (2005). Antispasmodic activity of fractions and cynaropicrin from *Cynara scolymus* on guinea-pig ileum. *Biol Pharm Bull* 28 (5): 902-4.

Zhu, X.F. *et al.* (2005). Antifungal activity of *Cynara scolymus* L. extracts. *Fitotarapia* 76 (1): 108-11.

Luppatelli, G. *et al.* (2004). Artichoke juice improves endothelial function in hyperlipemia. *Life Sci* 76 (7): 775-82.

Bundy, R. *et al.* (2004). Artichoke leaf extract reduces symptoms of irritable bowel syndrome and improves quality of life in otherwise healthy volunteers suffering from concomitant dyspepsia: a subset analysis. *J Alter Complement Med* 10 (4): 667-9.

GOAT'S RUE

Botanical Name: *Galega officinalis* L.

Natural Order: Fabaceae.

Active Ingredients: Alkaloid, flavonoids (Champavier *et al* 2000), guanidine derivatives: galegine (Barger and White

1923), 4-hydroxygalegine, glucoside: galeuteolin (Barger and White 1923).

Non-Clinical Studies: Anibacterial (Pundarikakshudu *et al* 2001), anti-obesity (Palit *et al* 1999), anti-platelet (Atanasov and Tchorbanov 2002) and hypoglycemic.

BIBLIOGRAPHY

Atanasov, T.A. and Tchorbanov, (2002). Anti-Platelet Fraction from *Galega officinalis* L. Inhibits Platelet Aggregation. *Journal of Medicinal Food* 5 (4): 229-234.

Barger, G. and White, F.D. (1923). Galeuteolin, a new glucoside from *Galega officinalis. Biochem J* 17 (6): 836–838.

Barger, G. and White, F.D. (1923). The Constitution of Galegine. *Biochem J* 17 (6): 827–835.

Champavier, Y. *et al.* (2000). Acetylated and non-acetylated flavonol triglycosides from *Galega officinalis. Chem Pham Bull* 48 (2): 281-282.

Palit, P. *et al.* (1999). Novel Weight-Reducing Activity of *Galega officinalis* in Mice. *Journal of Pharmacy and Pharmacology* 11 (1): 313-319.

Pundarikakshudu, K. *et al.* (2001). Anti-bacterial activity of *Galega officinalis* L. (Goat's Rue). *J Ethnopharmacol* 77(1): 111-112.

GOKSHURA

Botanical Name: *Tribulus terrestris* Linn.

Natural Order: Zygophyllaceae

Active Ingredients: Steroid saponins: disogenin, gitogenin, chlorogenin, ruscogenin, dioscin, terrestrinins A and B, tribulosin, tribol (Gheorghiu and Ioneacu-Matiu 1968; Gheorghiu and Ioneacu-Matiu 1968; Iskendrow 1970; Preepelista and Kintya 1974; Purushothaman, Chandrasekharan and Balakrishna 1976; Deepak *et al* 2002; Huang *et al* 2003; Conard *et al* 2004), flavonoids, alkaloids : harmine and harmaline (Borokowski and Lutomski 1960), cinnamic amide derivatives: terrestriamide, 7-methylhydroindanone-1, tribulusamide A and B (Ren *et al* 1994), lignanamides: tribulusamide A and B (Li *et al* 1998).

Non-Clinical Studies: Anticancer (Yang, Qu and Sun 2005), antihypertensive (Phillips, Mathew and Oriowo 2005), antifungal (Zhang *et al* 2005), anti-urolithiasis (Anand *et al* 1994; Satish, Periasamy and Namasivayam 1996), anthelmintic (Deepak *et al* 2002) aphrodisiac (Adikan *et al* 2000; Gauthaman, Adaikan and Prasad 2002), hepatoprotective (Li *et al* 1998), hyperlipidemia (Chu *et al* 2003), hypoglycemic (Li *et al* 2002), diuretic (Al-Ali *et al* 2003) and vasodilator (Phillips Mathew and Oriowo 2005).

Clinical Studies: A study conducted in twenty-one healthy young 20-36 years males demonstrated that steroid saponins present in *Tribulus terrestris* lacks direct or indirect androgen-increasing properties (Neychev and Mitev 2005).

Adverse Drug Reactions: Gynaecomastia (Jammel *et al* 2004).

Toxicity: Hepatotoxic in rats (Anonymous 1963; Anonymous 1964).

BIBLIOGRAPHY

Adikan, P.G *et al.* (2000).Proerectile pharmacological effects of *Tribulus terrestris* extract on the rabbit corpus cavernosum, *Ann Acad Med* 29 (1): 22-26.

Al-Ali, M. *et al.* (2003). *Tribulus terrestris*: preliminary study of its diuretic and contractile effects and comparison with *Zea mays*. *J Ethnopharmacology* 85 (2-3): 257-260.

Anonymous (1963). Toxicity of *Gokharu* seeds (*Tribulus terrestris*), *Indian Forester* 89 (12): 78.

Anonymous (1964). Toxicity of *Tribulus terrestris*. *Agric. Res.* 4 (1): 54.

Anand, R. *et al.* (1994). Activity of certain fractions of *Tribulus terrestris* fruits against experimentally induced urolithasis in rats. *Indian J Exp Biol* 32 (8): 548-552.

Borokowski, B. and Lutomski, J. (1960). Chromatographic examination of the alkaloid fraction from the herb and seeds of *Tribulus terrestris*. *Chem Abstr* 55 (24): 2519c.

Chakraborty, B and Neogi, N.C. (1978). Pharmacological properties of *Tribulus terrestris* Linn. *Indian J Pharm Sci* 40 (2) 50-52.

Chu, S. *et al.* (2003). Effect of saponin from *Tribulus terrestris* on hyperlipidemia. *Zhong Yo Cai* 26 (5): 341-344.

Conard, J. *et al.* (2004). A novel furostanol saponin from *Tribulus terrestris* of Bulgarian origin. *Fitoterapia* 75 (2): 117-122.

Deepak, M. *et al.* (2002). Tribulosin and beta-sitosterol-D-glucoside, the anthelmintic principles of *Tribulus terrestris*. *Phytomedicine* 9(8): 753-756.

Gauthaman, K., Adaikan, P.G. and Prasad, R.N. (2002). Aphrodisiac properties of *Tribulus terrestris* extract (Protodioscin) in normal and castrated rats. *Life Sci* 71 (12): 1385-1396.

Gheorghiu, A and Ioneacu-Matiu, E (1968). Presence of chlorogenin with disogenin and gitogenin in *Tribulus terrestris*, *Ann Pharm Fr* 26 (12): 745-748.

Gheorghiu, A and Ioneacu-Matiu, E (1968). Steroid sapogenins, presence of chlorogenin in *Tribulus terrestris Stud Cercet Biochim* 11(3): 269-273.

Huang, J.W. *et al.* (2003). Terrestrinins A and B, two new steroid saponins from *Tribulus terrestris*. *J Asain Nat Prod Res* 5 (4): 285-290.

Iskendrow, G.B. (1970). Steroidal sapogenins from *Tribulus terrestris*, *Khim Prir Sodin* 6 (4): 488-489.

Jammel, J.K *et al.* (2004). Gynaecomastia and the plant product "*Tribulis terrestris*". *Breast* 13 (5): 428-430.

Kumari, G.S. and Iyer, G.Y. (1967). Preliminary studies on the diuretic effects of *Hygrophila spinosa* and *Tribulus terrestris*, *Indian J Med Res* 55 (7): 714-716.

Li, J.X. *et al.* (1998). Tribulusamide A and B, new hepatoprotective lignanamides from the fruits of Tribulus terrestris: indications of cytoprotective activity in murine hepatocyte culture. *Planta Med* 64 (7): 628-31.

Li, M. *et al.* (2002). Hypoglycemic effect of saponin from *Tribulus terrestris*. *Zhong Yao Cai* 25 (6): 420-422.

Neychev, V.K. and Mitev, V.I. (2005). The aphrodisiac herb *Tribulus terrestris* does not influence the androgen production in young men. *J Ethnopharmacol* 101 (1-3):319-323.

Phillips, O.A., Mathew, K.T. and Oriowo, M.A. (2006). Antihypertensive and vasodilator effects of methanolic and aqueous extracts of *Tribulus terrestris* in rats. *J Ethnopharmacol*. 104(3): 351-355.

Preepelista, E.D. and Kintya, P.K. (1974). Dioscin, a steroid glycoside from *Tribulus terrestris* L. *Biol Abstr* 61, 9581.

Purushothaman, K.K.; Chandrasekharan, S. and Balakrishna K. (1976). Occurrence of neotigogenin and hecogenin in *Tribulus terrestris* Linn. *J Res Indian Med Yoga and Homeo* 11 (1): 121-125.

Ren, Y.J. *et al.* (1994). Isolation and identification of a new derivative of cinnamic amide from *Tribulus terrestris. Acta Pharm Sinica* 29(3): 204-206.

Satish, S.; Periasamy, P. and Namasivayam, A. (1996). Effect of *Tribulus terrestris* on experimental urolithiasis induced by ethylene glycol in albino rats, *Pharm Sci* 2 (9): 437-439.

Yang, H.J., Qu, W.J. and Sun, B. (2005). Experimental study of saponins from *Tribulus terrestris* on renal carcinoma cell line. *Zhongguo Zhong Yao Za Zhi.* 30 (16): 1271-1274.

Zhang, J.D. *et al.* (2005). *In vitro* and *in vivo* antifungal activities of the eight steroid saponins from *Tribulus terrestris* L. with potent activity against fluconazole-resistant fungal. *Biol Pharm Bull* 28(12): 2211-2215.

GOLDEN SEAL

Botanical Name: *Hydrastis canadensis* Linn.

Natural Order: Ranunculaceae.

Active Ingredients: Alkaloids: hydrastinine (Weber *et al* 2003), berberastine (Weber *et al* 2003), tetrahydroberberastine (Weber *et al* 2003), canadaline (Scazocchio *et al* 2001; Weber *et al* 2003), berberine (Scazocchio *et al* 2001; Inbaraj *et al* 2001; Weber *et al* 2003), hydrastine, (Torss *et al* 1951; Scazocchio *et al* 2001; Weber *et al* 2003), coptisine, jatrorrhizine (Weber *et al* 2003) and canadine (Scazocchio *et al* 2001; Weber *et al* 2003).

Non-Clinical Studies: Antibacterial (Scazocchio *et al* 2001), phototoxic (Inbaraj *et al* 2001), relaxant on isolated guinea pig trachea (Haq *et al* 2000).

BIBLIOGRAPHY

Haq, A *et al.* (2000). Relaxant effects of *Hydrastis canadensis* L. and its major alkaloids on guinea pig isolated trachea. *Pharmacol Toxicol* 87 (5): 218-222.

Inbaraj, J.J. *et al.* (2001). Photochemistry and photocytotoxicity of alkaloids from Goldenseal (*Hydrastis canadensis* L.) 1. Berberine. *Chem Res Toxicol* 14 (11): 1529-1534.

Scazocchio, F. *et al.* (2001). Antibacterial activity of *Hydrastis canadensis* extract and its major isolated alkaloids. *Planta Med* 67 (6): 561-564.

Torss, C.J. *et al.* (1951). Note on the hydrastine content in cultivated *Hydrastis canadensis* L. *Farmacognosia* 11 (22): 425-431.

Weber, H.A. *et al.* (2003). Chemical comparison of goldenseal (*Hydrastis canadensis* L.) root powder from three commercial suppliers. *J Agr Food Chem* 51 (25): 7352-7358.

GREATER CELANDINE

Botanical Name: *Chelidonium majus* L.

Natural Order: Papaveraceae

Active Constituents: Alkaloids: berberine, coptisine, cryptopine, chelidonine, chelerythrine, protopine, sanguinarine, stylopine (Saqlam and Arar 2003), bitter constituent, proteolytic enzyme and thiophosphoric acid triamide.

Non-Clinical Studies: Cytotoxic (Anon: PDR for Herbal Medicines; Sokoloff *et al* 1964: Shi 1992; Hüsniye Sağlam 2002; Saqlam and Arar 2003; Uglyanitsa *et al* 2002), oxytocic (Anon: PDR for Herbal Medicines), positively ionotropic (Anon: PDR for Herbal Medicines), chemo preventive (Kim *et al* 1997), antimutagenic (Shi 1992), antiviral (Lozjuk Lisnyak and Lozjuk 1996), cholretic (Vahlensiek*et al* 1995), anti-inflammatory (Vavreckova *et al* 1996), hepatoprotective (Mitra *et al* 1992), cholecystokinetic (Hriscu, Galesanu and Moisa 1980), antispasmodic and relaxant (Hiller, Ghorbani and Schilcher 1998).

Clinical Studies

Dyspepsia

A double-blind trial found that a standardized extract of greater celandine could relieve symptoms of dyspepsia or indigestion significantly better than a placebo. The trial used an extract standardized to 4 mg of chelidonine per capsule and gave

1-2 tablets three times daily for six weeks. An earlier, preliminary trial also found the same extract reduced symptoms in people with indigestion (Ritter and Schatton 1993).

Prostate carcinoma

90 patients with histologically proven unresectable pancreatic cancer were randomized in a monocentric, controlled, randomized study. Patients in arm A received 1000 mg gemcitabine/m^2, those in arm B received 20 mg NSC-631570, and those in arm C received 1000 mg gemcitabine/m^2 followed by 20 mg NSC-631570 weekly. End point of the study was overall survival (Gansauge *et al* 2002).

Pharmacokinetics: Following administration, Ukrain (thiophosphoric acid derivative of *Chelidonium majus* L. alkaloids) distributes rapidly throughout the body, including the brain and central nervous system. Most of the compound remains unmodified and is excreted through the kidneys. Ukrain administered to rats intraperitoneally at a dose of 28 mg/kg revealed rapid distribution into the plasma and a biologic half-life of approximately 60 minutes. Acute toxicity studies indicate that LD50 dose for intraperitoneally administered Ukrain in mice and rats is approximately 280 mg/kg. Maximum tolerated intravenous dose was 3.5 mg/kg in rats and 0.35 mg/kg in rabbits. Chronic toxicity studies in mice and rats with intraperitoneal administration revealed no apparent toxic effects (Jagiello-Wojtowicz *et al* 1998).

Adverse Drug Reation: Contact dermatitis (Etxenagusia *et al* 2000).

Toxicity: Respiratory depressant (Barreto *et al* 2003), hepatotoxic (Benninger*et al* 1999). LD_{50} for ethyl acetate, methanol, aqueous and ethanol extracts: 152.42, 395.09, 1070.76, 1172.77, respectively (Hüsniye 2002).

BIBLIOGRAPHY

Anon (2000). PDR for Herbal Medicines. Medical Economics Company Inc, Montvale, NJ: 812-813.

Barreto, M.C. *et al.* (2003). Inhibition of mouse liver respiration by *Chelidonium majus* isoquinoline alkaloids. *Toxicol Lett* 146 (1): 37-47.

Benninger, J. *et al.* (1999). Acute hepatitis induced by greater celandine (*Chelidonium majus*). *Gastroenterol* 117 (5): 1234-7

Bone, K (1996). Chelidonium--A medicinal poppy. *Medi Herb Professional Newsletter* 49: 1–3.

Etxenagusia, M.A., *et al.* (2000). Contact dermatitis from *Chelidonium majus* (greater celandine). *Contact Derm* 43:47.

Gansauge, F. *et al.* (2002). Results of phase 2 trial. NSC-631570 (Ukrain) in the palliative treatment of pancreatic cancer. *Langenbeck's Arch Surg* 386: 570-4.

Hiller, K.O.; Ghorbani, M. and Schilcher, H. (1998). Antispasmodic and relaxant activity of chelidonine, protopine, coptisine, and *Chelidonium majus* extracts on isolated guinea-pig ileum. *Planta Med* 64: 758–60.

Hriscu, A.; Galesanu, M.R. and Moisa, L. (1980). Cholecystokinetic action of an alkaloid extract of *Chelidonium majus*. *Rev Med Chir Soc Med Nat Lasi*; 84:559–61.

Hüsniye Sağlam (2002). Cytotoxic properties of *Chelidonium majus* L. Sağlam –140-142. 140 Proceedings of ICNP-2002–Trabzon / TÜRKİYE.

Jagiello-Wojtowicz, E. *et al.* (1998). Preliminary pharmacokinetic studies of Ukrain in rats. *Drugs Exp Clin Res* 24: 309-11.

Kim, D.J. *et al.* (1997). Potential preventive effects of *Chelidonium majus* L. (Papaveraceae) herb extract on glandular stomach tumor develo-pment in rats treated with N-methyl-N'-nitro-Nnitrosoguanidine (MNNG) and hypertonic sodium chloride. *Cancer Lett* 112 (2): 203-8.

Lozjuk, R.M.; Lisnyak, O.I. and Lozjuk, L.V. (1996). Theoretical grounds and experimental confirmation of the antiviral effect of the preparation Ukrain. *Drugs Exp Clin Res* 22 (3-5): 213-7

Mitra, S. *et al.* (1992). Antihepatotoxic activity of Chelidonium majus. *Int J Pharmacognosy* 30: 125–8.

Ritter, R. and Schatton, W.F.H. (1993). Clinical trial on standardized celandine extract in patients with functional epigastric complaints: Results of placebo-controlled double-blind trial. *Comp Ther Med* 1: 189–93.

Saqlam, H. and Arar, G. (2003). Cytotoxic activity and quality control determinations on *Chelidonium majus*. *Fitoterapia* 74 (1-2): 127-129.

Shi, G.Z. (1992). Blockage of *Glychyrrhiza uralensis* and *Chelidonium majus* in MNNG induced cancer and mutagenesis. *Chung Hua Yu Fang I Hsueh Tsa Chih* 26 (3): 165-7.

Sokoloff, B. *et al.* (1964). The antitumor factors present in *Chelidonium majus* L. I. Chelidonine and protopine. *Growth* 28: 225–31.

Uglyanitsa, K.N., *et al.* (2002). Ukrain: a novel antitumor drug. *Drugs Exp Clin Res* 56: 347-56.

Vahlensiek, U. *et al.* (1995). The effect of *Chelidonium majus* herb extract on the choleresis in the isolated perfused rat liver. *Planta Med* 61(3): 267-270.

Vavreckova, C. *et al.* (1996). Benzophenanthridine alkaloids of *Chelidonium majus*: I. Inhibition of 5- and 12- lipoxygenase by a non-redox mechanism. *Planta Med* 62 (5): 397-401.

GUDUCHI

Botanical Name: *Tinospora cordifolia* (Wiild.) Miers ex. Hook. f. and Thomas.

Natural Order: Menispermaceae.

Active Ingredients: Alkaloid: berberine, glucoside: giloin, tinosporinine, starch, diterpenoid furanolactone: tinosporide, arabinogalactan (Chintalwar *et al* 1999), polysaccharide (Jahfar 2003).

Non-Clinical Studies: Anti-allergic, immunomodulator (Chintalwar *et al* 1999; Singh, Singh and Shrivastva 2004), colony stimulating (Thatte, Rao and Dahanukar 1994), hepatoprotective, hypolipidemic (Mainzen *et al* 1999), antioxidant (Goel *et al* 2002), antitumor (Singh, Singh and Shrivastva 2004), antiulcer (Sharma *et al* 1995) anti-stress (Sharma *et al* 1995).

BIBLIOGRAPHY

Chintalwar, G. *et al.* (1999). An immunologically active arabinogalactan from *Tinospora cordifolia*. *Phytochemistry* 52 (6): 1089-93.

Goel, H.C. *et al.* (2002). Free radical scavenging and metal chelation by *Tinospora cordifolia*, a possible role in radioprotection. *Indian J Exp Biol* 40 (6): 727-34.

Jahfar, M. (2003). Glycosyl composition of polysaccharide from *Tinospora cordifolia*. *Acta Pharm* 53 (1): 65-9.

Thatte, U.M.; Rao, S.G. and Dahanukar, S.A. (1994). *Tinospora cordifolia* induces colony stimulating activity in serum. *J Postgrad Med* 40:202-3

Mainzen, S. *et al.* (1999). Hypolipidaemic action of *Tinospora cordifolia* roots in alloxan diabetic rats. *J Ethnopharmacol* 64 (1): 53-7.

Singh, N.; Singh, S.M. and Shrivastva, P. (2004). Immunomodulatory and antitumor actions of medicinal plant *Tinospora cordifolia* are mediated through activation of tumor-associated macrophages. *Immunopharmacol Immunotoxicol* 26 (1); 145-62.

Sharma, D.N.K. *et al.* (1995). Antiulcer activity of *Tinsopora cordifolia* Miers. and *Centella asiatica* Linn. extracts, *Phytotherapy Research* 9 (8): 589-590.

Sharma, D.N.K. *et al.* (1995). Antistress activity of *Tinsopora cordifolia* Miers. and *Centella asiatica* Linn. extracts, *Phytotherapy Research* 10 (2): 181-183.

GUGGUL

Botanical Name: *Commiphora mukul* Roxb. ex Colebr. Syn *Boswellia glabra* (Roxb.) Bennet, *Boswellia glabra* Roxb.

Natural Order: Burseraceae.

English Name: Indian Bdellium tree or Gum guggal.

Active Ingredients: Guggulsterones (E and Z), gugulsterols (Patil *et al* 1972; Roy *et al* 1989), lignan: commiphorin (Sultana *et al* 2004) and fatty acid ester: commiphotetrol (Sultana *et al* 2004).

Non-Clinical Studies: Antibacterial (Sultana *et al* 2004), antidiabetic (Subramaniam *et al* 2001), anti-inflammatory (Sharma and Sharma 1977; Dahanukar *et al* 1983; Satyavati, 1991; Duwiejua *et al* 1993), antioxidant (Singh, Niaz and Ghosh 1994) and hypolipidemic (Satyavati 1966; Satyavati Dwarakanath, Tripathi 1969; Malhotra, Ahuja and Sundaram 1977; Kuppurajan *et al* 1978; Singh, Niaz and Ghosh 1994; Verma and Bordia 1998; Szapary *et al* 2003).

According to a recent study, guggulsterone activates nuclear receptors such as estrogen receptor a, pregnane X receptor and progesterone receptor. The study also showed that activation of pregnane receptor by guggulsterone led to induction of CYP3A genes *in vitro* (Dan *et al* 2004).

Clinical Studies

Acne vulgaris

A study described efficacy of *Commiphora mukul* in treating acne vulgaris similar to tetracycliclins (Thappa and Dogra 1994). However results are inconclusive.

Hyperlipidemia

A. This randomized, double-blind study divided 61 patients aged 25-65 years with serum cholesterol levels >200 mg/dl into two groups. Throughout the study, both groups were advised to eat at least 400 g/day of fruits and vegetables and had other dietary restrictions aimed to lower their cholesterol levels. One group was given 50 mg guggulipid BID for 24 weeks, followed by a 12 week washout period. The second group was given a placebo in place of the guggulipid. While the diet alone caused a significant decrease in total cholesterol, LDL and triglycerides in both groups, those treated with the guggulipid adjunct had significant further reduction of nearly twice as much as with the diet alone.

   The washout period results further pointed to guggulipid as the producer of the noted effect (Singh *et al* 1994). This clinical trial divided 120 patients into obese, hyper-cholesterolemic or hyperlipemic groups, which were further divided into 21-day treatment arms of: (1) 2 g three TID 'purified' guggulu (according to Ayurvedic texts), (2) 500 mg BID Fraction 'A' of petroleum ether extract of gum guggulu, (3) placebo or (4) 500 mg TID clofibrate. Significant reductions in Serum cholesterol and total lipids were seen in the Fraction 'A' treatment arm in the hypercholesterolemic group. The reduction was comparable to that seen with clofibrate. Other significant reductions occurred in other treatment arms, but larger, focused studies are needed to replicate this data (Kuppurajan *et al* 1978).

B. Clinical trial of 103 patients with hypercholesterolemia. Patients were randomly chosen to receive 1 gram guggulipid, 2 grams guggulipid or placebo three times daily for eight weeks. Patients in both of the guggulipid groups failed to show any lowering of total cholesterol, HDL cholesterol, triglycerides or VLDL cholesterol. Furthermore, patients treated with guggulipid were found to have statistically significant increases in LDL cholesterol levels when compared with those in the placebo group. Researchers suggest that the diet or genetic makeup of patients in previous Indian trials of guggulipid may have accounted for some of the perceived cholesterol-lowering effects. Researchers also suggest that previous studies did not use the standard ultaracentrifugation technique to directly measure levels of LDL cholesterol (Szapary *et al* 2003).

Obesity

A small double-blind, placebo controlled study failed to prove efficacy of a standardized guggulsterone phosphate supplement in obese patients (Antonio *et al* 1999).

Osteoarthritis

A study involving a total of 30 patients with osteoarthritis of knee found significant reductions in swelling and pain over the course of 2 months with *Commiphora mukul*. The drug was administered in capsule form (500 mg concentrated exact delivered thrice-a-day) along with food. Based on evidence it was concluded that there were no side effects reported during the trial. *Commiphora mukul* appears to be a relatively safe and effective supplement to reduce symptoms of osteoarthritis (Singh *et al* 2003).

Rheumatoid arthritis

Clinical trials with purified *Commiphora mukul* has been carried out in 35 patients of rheumatoid arthritis in order to assess its anti rheumatic activity, dose requirement, resistance development and side effects on haematology. From the results obtained it has been indicated that *Commiphora mukul* acts as an analgesic agent without any toxic or side effect (Vyas *et al* 1987).

Adverse Drug Reactions

Diarrhea (Singh, Niaz and Ghosh 1994), hypersensitivity (Szapary*et al* 2003), gastritis, mouth ulcer and menstrual abnormalities.

Drug-Interactions Anticoagulants, thyroid medications (Singh, Niaz and Ghosh 1994) and calcium channel blockers (diltiazem).

BIBLIOGRAPHY

Antonio, J., *et al.* (1999). Effects of a standardized guggulsterone phosphate supplement on body composition in overweight adults: a pilot study. *Curr Ther Res* 60: 220–227.

Dan, E., *et al.* (2004). Guggulsterone activates multiple nuclear receptors and induces CYP3A gene expression through the pregnane X receptor. *JPET* 310: 528-535.

Duwiejua, M. *et al.* (1993). Anti-Inflammatory Activity of Resins from Some Species of the Plant Family Burseraceae. *Planta Medica* 59 (1): 12-6.

Kuppurajan, K., *et al.* (1978). Effect of guggulu (*Commiphora mukul* Engl.) on serum lipids in obese, hypercholesterolemic and hyperlipemic cases. *J Assoc Physicians India* 26: 367-73.

Malhotra, S.C., Ahuja, M.M. and Sundaram, K.R. (1977). Long term clinical studies on the hypolipidaemic effect of Com*miphora mukul* (Guggulu) and clofibrate. *Indian J Med Res* 65:390-5.

Satyavati, G.V. (1966). Effect of an indigeneous drug on disorders of lipid metabolism with special reference to atherosclerosis and obesity (Medoroga). M.D. Thesis (Doctor of Ayurvedic Medicine), Banaras Hindu University, Varanasi, 1966.

Satyavati, G.V. (1988). Gum guggul (*Commiphora mukul*)-The success story of an ancient insight leading to a modern discovery. *Ind J Med Res* 87: 327.

Satyavati, G.V., Dwarakanath, C. and Tripathi, S.N. (1969). Experimental studies on the hypocholesterolemic effect of *Commiphora mukul.* Engl. (Guggul). *Indian J Med Res* 57: 1950-62.

Singh, B.B., *et al.* (2003). The effectiveness of *Commiphora mukul* for osteoarthritis of the knee: an outcomes study. *Altern Ther Health Med.* 9 (3): 74-9.

Singh, R.B., Niaz, M.A. and Ghosh, S. (1994). Hypolipidemic and antioxidant effects of *Commiphora mukul* as an adjunct to dietary therapy in patients with hypercholesterolemia. *Cardiovasc Drugs Ther* 8: 659-64.

Scartezzini, P. and Speroni, E. (2000). Review on some plants of Indian traditional medicine with antioxidant activity. *J Ethnopharmacol* 71: 23-43.

Sharma, J.N. and Sharma, J.N. (1977). Comparison of the anti-inflammatory activity of *Commiphora mukul* (an indigenousdrug) with those of phenylbutazone and ibuprofen in experimental arthritis induced by mycobacterial adjuvant *Arzneimittelforschung* 27 (7): 1455-7.

Subramaniam, A., *et al.* (2001). Guggul lipid reduces insulin resistance and body weight gain in C57B1/6 lep/lep mice [abstract]. *Int J Obes Relat Metab Disord* 25 (2): S24.

Sultana, N, *et al.* (2005). Studies on the constituents of *Commiphora mukul*. Z *Naturforsch* 60b: 1202 – 1206.

Szapary, P.O. *et al.* (2003). Guggulipid for the treatment of hyper cholesterolemia: a randomized controlled trial. *JAMA* 290: 765-72.

Thappa, D.M. and Dogra, J. (1994). Nodulocystic acne: oral gugulipid versus tetracycline. *J Dermatol* 21:729–731.

Urizar, N.L. *et al.* (2002). A natural product that lowers cholesterol as an antagonist ligand for FXR. *Science* 296 (5573): 1703-6.

Verma S.K. and Bordia, A. (1988). Effect of *Commiphora mukul* (gum guggulu) in patients of hyperlipidemia with special reference to HDL-cholesterol. *Indian J Med Res* 87: 356-60.

HARITAKI

Botanical Name: *Terminulia chebula* Retz.

Natural Order: Combretaceae

Active Ingredients: Chebulinic acid (Barat 1953), chebulagic acid (Schmidt and Nieswandt 1950), ellagic acid (Schmidt and Nieswandt 1948), tannic acid, terchebin (Okuda *et al* 1980), tetrachebulin (Lin *et al* 1995), chebulin (Gaind and Saini 1965), Pentacyclic triterpene: 2-α-Hydroxymicromeric acid (Singh 1990), chebupentol (Chuttani *et al* 2003).

Non-Clinical Studies: Antibacterial (Malekzadeh *et al* 2001): anticancer (Antimicrobial, Nandy *et al* 1997), antifungal (Srivastava and Srivastava 2005), anti-mutagenic (Grover and Sarojbala 1992; Kaur *et al* 1998), antispasmodic, cytotoxic (Lee *et al* 1995), cardio-tonic (Reddy *et al* 1994); hypolipidemic

(Thakur *et al* 1985), hypotensive (Guan *et al* 1996), immuno-supresasant (*et al* 1997), prokinetic (Tamhane *et al* 1997).

BIBLIOGRAPHY

Barat, S.K. (1953). A note on tannage with chebulinic acid. *J. Indian Chem. Soc* 16 (4): 217.

Chuttani, K., *et al.* (2003). Radiolabelling and biological Evaluation of a non-peptidic compound from *Terminalia chebula* (Harar) for CCK expressing tumours. *IJNM* 18 (1 and 2): 19-24.

Das, B.M., Barat, S.K. and Scaria, K.J. (1956). Nature of tannins in *Terminalia chebula*, Proc. 43rd Indian Sci. Congr., Part-III, Abstr: 85.

Gaind, K.N. and Saini, T.S. (1965). Identification of purgative principle of *Terminalia chebula* Retz. *Indian J Pharm* 30 (10): 233-234.

Grover, N.N. and Sarojbala (1992). Antimutagenic activity of *Terminalia chebula* in *Salmonella typhimurium*, *Indian J Expt Biol* 30(4): 339-341.

Guan, Y., *et al.* (1996). *In vitro* inhibitory effects of chebulinic acid on the contractile responses of cardiovascular muscles. *Clin Exp Pharmacol Physiol* 23 (8): 747-50.

Hamada, S., *et al.* (1997). Immunosuppressive effects of gallic acid and chebulagic acid on CTL-mediated cytotoxicity. *Biol Pharm Bull* 20 (9): 1017-9.

Kaur, S., *et al.* (1998). Antimutagenicity of hydrolysable tannins from *Terminalia chebula* in *Salmonella typhimurium*. *Mutat Res* 419 (1-3): 169-179.

Lee, S.H., *et al.* (1995). Hydrolysable tannins and related compounds having cytotoxic activity from the fruits. *Arch Pharmacol Res* 18 (2): 118-120.

Lin, T., *et al.* (1995). Tannins and related compounds CII., Structures of terchebulin, an ellagitannin having a novel tetraphenyl carboxylicac id moiety and biogenetically related tannins from *Terminalia chebula* Retz., *Chem Pharm Bull* 38 (11): 3004-3008.

Malekzadeh, F. *et al.* (2001). Antibacterial activity of black myrobalan (*Terminalia chebula* Retz) against *Helicobacter pylori*. *Int J Antimicrob Agents* 18 (1): 85-8.

Okuda, T., *et al.* (1980). Hydrolysable tannins having enatiomeric dehydrohexahydrodiphenyl group: Revised structure of terchebin and structure of granatin B. *Tetrahedron Letters* 21 (45): 4361-4364.

Reddy, B.M., *et al.,* (1994). Cardio tonic activity of the fruit of *Terminalia chebula. Int J Pharmacog* 32 (4) 352-356.

Schmidt, O.T. and Nieswandt, W. (1948). Crystalline tanning agent containing ellagic acid from myrobalan, *Naturwissenschaften* 35 91.

Schmidt, O.T. and Nieswandt, W. (1950). Natural tannins. III: Chebulagic acid, a crystalline ellagitannin from myrobalans. *Ann* 568: 165-167.

Singh, C. (1990). 2-α-Hydroxymicromeric acid, a pentacyclic tritepene from *Terminalia chebula, Phytochemistry* 29 (7): 2348-2350.

Srivastava, S.K. and Srivastava, S.D. (2005). New Biologically active Constituents from *Terminalia chebula* Stem Bark. *Cheminform* 36 (17): 25.

Tamhane, M.D. *et al.* (1997). Effect of oral administration of *Terminalia chebula* on gastric emptying: an experimental study. *J Postgrad* Med 43: 12-3.

Thakur, C.P. *et al.* (1985). The Ayurvedic Medicines *Haritaki, Amala* and *Bahira* reduce cholesterol-induced atherosclerosis in rabbits. *Int J Cardiol* 21 (2) 134-135.

HAWTHORN

Botanical Name: *Crataegus oxyacantha* Linn.

Natural Order: Rosaceae.

Active Ingredients: Amygladin, alkaloid: crataegin, flavonoids: rutin (Rehwald *et al* 1993), proanthocyanidins, pectin, cardiotonic amines: phenylethlamine, tyramine, isobutylamine, o-methoxyphenylethylamine.

Non-Clinical Studies: Anti-arrhythmic (Krzeminski and Chatterjee 1993; Shuessler *et al* 1993), coronary vasodilator (Gildor 1988; Shuessler *et al* 1993), antioxidant (Bahorun *et al* 1994; Bahorun *et al* 1996), anti-hyperglycemic (Jaoud *et al* 2003), hypolipidemic (Shanti *et al* 1994).

Clinical Studies

Congestive cardiac failure

A total of 143 patients (72 men, 71 women, mean age of 64.8 (8.0 years) were recruited and treated with 3 times 30 drops of the extract (n = 69) or placebo (n = 74) for 8 weeks. The primary variable for the evaluation of efficacy was the change in exercise tolerance determined with bicycle exercise testing, secondary variables included the blood pressure-heart rate product (BHP). Subjective cardiac symptoms at rest and at higher levels of exertion were assessed by the patient on a categorical rating scale. An overall assessment of efficacy at the final visit was provided by the patient and the investigator.

In the ITT population there was a significant increase in exercise tolerance in both groups between visit 1 and visit 3. The difference between the treatment groups was 8.3 watts in favour of the standardised extract of fresh Crataegus berries ($p = 0.045$). The result is confirmed in the PP population ($p = 0.047$). Changes in BHP at 50 watts and at comparable maximum load were in favour of Crataegus extract but the results are not statistically significant.

The subjective assessment of cardiac symptoms at rest and at higher levels of exertion did not change significantly and the patient and investigator overall assessment of efficacy were similar for the two groups. The medication was well tolerated and had a high level of patient acceptability.

The significant improvement, due to the fact that dyspnoea and fatigue do not occur until a significantly higher wattage has been reached in the bicycle exercise testing allows the conclusion that the recruited NYHA II patients may expect an improvement in their heart failure condition under long term therapy with the standardised extract of fresh Crataegus berries (Denerging *et al* (2003).

Drug Interactions: No drug interaction with digoxin (Tankanow *et al* 2003).

Toxicity: LD50 of water extract in mice: 13.5 g/kg body weight (Schlegelmilch and Heywood 1994). Acute oral toxicity: 18.5-33.8 ml/kg and 6 g/kg body weight.

BIBLIOGRAPHY

Bahorun, T. *et al.* (1994). Antioxidant activities of *Cratageus monogyna* extracts. *Planta Med* 60 (4): 323-328.

Bahorun, T. *et al.* (1996). Oxygen species scavenging activity of phenolics, fresh plant organ and pharmacological preparations. *Arzneim Forsch* 46 (11): 1086-1089.

Denerging, F.H. *et al.,* (2003). A randomised double blind placebo controlled clinical trial of a standardised extract of fresh Crataegus berries (Crataegisan) in the treatment of patients with congestive heart failure NYHA II. *Phytomedicine* 10 (5): 363-9.

Gildor, A. (1988). *Crataegus oxyacantha* and heart failure. *Circulation* 98 (19): 2098.

Jaoud, H *et al.* (2003). Hawthorn evokes a potent anti-hyperglycemic capacity in streptozotocin-induced diabetic rats. *J Herb Pharmacother* 3 (2): 19-29.

Krzeminski, T. and Chatterjee, S.S. (1993). Ischemia and early reperfusion induced arrhythmias, beneficial effects of an extract of *Crataegus oxycantha* L. *Pharm Pharmacol Lett* 3: 45-48.

Rehwald, A. *et al.,* (1993). HPLC analysis of the flavonoids of Cratageus folium cum flore. *Planta Med* 59 (7): 28.

Schlegelmilch, R. and Heywood, R. (1994). Toxicity of Crataegus extract. *J Am Coll Toxicol* 13: 103-111.

Shanti, S. *et al.,* (1994). Hypolipidemic activity of tincture of Crataegus in rats. *Indian J Biochem Biophys* 31(2):143-6.

Shuessler M. *et al.,* (1993). Cardiac effects of flavnoids from Cratageus species. *Planta Med* 59 (7): 88.

Tankanow, R. *et al.,* (2003). Interaction study between digoxin and a preparation of hawthorn (*Crataegus oxyacantha*). *J Clin Pharmacol* 43 (6): 637-42.

HOREHOUND

Botanical Name: *Ballota nigra* Linn. ssp. *foetida* (Vis.) Hayek

Natural Order: Lamiaceae

Active Constituents: Diterpenes: 7-acetoxymarrubiin, ballonigrin, ballotinone, ballotenol, marrubiin, preleosibirin (Savona *et al* 1976; Savona *et al* 1977; Savona *et al* 1977; Bruno *et al* 1986; Seidel Bailleul and Tillequin 1998), acids and essential oil.

Non-Clinical Studies: Anti-arrythmic, antibacterial (Didry *et al* 1999), antioxidant (Daels-Rakotoarison *et al* 2000; Seidel; Bailleul and Tillequin 1998), sedative (Daels-Rakotoarison *et al* 2000), hypotensive and cholagouge.

BIBLIOGRAPHY

Bruno, M. *et al.* (1986). Preleosibirin, a prefuranic labdane diterpene from *Ballota nigra* ssp. *foetida. Phytochemistry* 25: 538-539.

Daels-Rakotoarison, D.A. *et al.* (2000). neurosedative and antioxidant activities of phenylpropanoids from *Ballota nigra. Arzneimittelforschung* 50 (1): 16-23.

Didry, N. *et al.* (1999). Isolation and antibacterial activity of phenylpropanoid derivatives from *Ballota nigra. Journal of Ethnopharmacology* 67 (2): 197-202.

Savona, G. *et al.* (1976). Structure of ballotinone, a diterpenoid from *Ballota nigra. J Chemical Society Perkin Trans* 1:1607-9.

Savona, G. *et al.* (1977). Structures of three new diterpenoids from Ballota species. *J Chemical Society Perkin Trans* 1: 322-324.

Savona, G. *et al.,* (1977). The structure of ballotenol, a new diterpenoid from *Ballota nigra. J Chemical Society Perkin Trans* 1: 497-9.

Seidel, V. *et al.* (2000). Phenylpropanoids from *Ballota nigra* L. inhibit *In vitro* LDL peroxidation. *Phytotherapy Research* 14 (2): 93-98.

Seidel, V.; Bailleul, F. and Tillequin, F. (1998). Diterpene and phenylpropanoid heteroside esters from *Ballota nigra* L. [Article in French] *Annales Pharmacie Francais* 56 (1): 31-35.

HORSE CHESTNUT

Botanical Name: *Aesculus hippocastanum* Linn.

Natural Order: Hippocastanaceae.

Active Constituents: Triterpene saponins or aescin or escin: contains β-aescin as major component and α-aescin: equal proportion of β-aescin and cryptoaescin (Patt and Wrinkler 1960; Yoshikawa *et al.* 1996; Yoshikawa *et al* 1998), hydroxycoumarins; aesculin, proanthocyanidins, flavonoids, carotenoid: asculaxanthin (Jozsef *et al* 1998).

Non-Clinical Studies: Anti-inflammatory (Aizawa *et al* 1974), cytotoxic (Konoshima and Lee 1986), hypoglycemic (Yoshikawa *et al* 1996), vasoconstrictor, venotomic (Vayssairat *et al* 1996; Annoni *et al* 1979).

Clinical Studies

Venous insufficiency

A. A clinical study of 22 patients with proven venous insufficiency measured by capillary filtration coefficient and intravascular volume of the lower legs showed that 1200 mg of HCSE, standardized to 50 mg escin per capsule, lowered capillary filtration by 22% in three hours (Bisler *et al* 1986).

B. In a randomized, placebo-controlled, parallel, double-blind study of 40 patients with venous edema in chronic deep vein incompetence, the edema-reducing effect of HCSE was deemed statistically significant. Measurements of leg circumference and other relevant indicators showed HCSE to be a safe and effective adjunct to compression therapy (Diehm *et al* 1992).

C. In a multi-center, placebo-controlled study on 60 patients (half receiving placebo), HCSE tablets (63–90 mg extract, standardized to 20 mg escin) given in a dosage of three tablets twice daily for six weeks, were effective in reducing edema in the ankles and the venous filling rate (Shah *et al* 1997).

Pharmacokinetics: Aescin adminstred orally has an absor-ption half-life of about 1 hour and an elimination half-life of about 20 hours (Schulz *et al* 1998).

Adverse Drug Reations: Contact dermatitis (Comaish and Kersey 1980).

Toxicity: Liver injury reported with intramuscular injection. It is nephrotoxic also. Aesculin effects bleeding time.

BIBLIOGRAPHY

Aizawa, X. *et al.* (1974). Anti-inflammatory action of Aescin (1, intra-venous injection). *Pharmacomertics* (Tokyo) 8: 211.

Annoni, F. *et al.,* (1979). Venotomic activity of Escin on the human saphenous vein. *Arzneim Forsch/Drug Res* 29: 672.

Bisler, H., R. Pfeifer, N. Kluken, P. Pauschinger. (1986). Wirkung von Rosskastaniensamenextrakt auf die traskapillare filtration bei chronischer venoser insuffizien [Effects of horse-chestnut seed extract on transcapillary filtration in chronic venous insufficiency]. *Dtsch Med Wochenschr* 111 (35): 1321–1329.

Comaish, J.S. and Kersey, P.J. (1980). Contact dermatitis to extract of horse chestnut (esculin). *Contact Dermatitis* 6 (2): 150-151.

Diehm, C. *et al.,* (1996). Comparison of leg compression stocking and oral horse-chestnut seed extract therapy in patients with chronic venous insufficiency. *Lancet* 347 (8997): 292–294.

Jozsef, D. *et al.,* (1998). Aesculaxanthin, a New Carotenoid Isolated from Pollens of *Aesculus hippocastanum. Helvetica Chimica Acta* 81 (1): 1815-1820.

Konoshima, T. and K.H. Lee. (1986). Antitumor agents, 82. Cytotoxic sapogenols from *Aesculus hippocastanum. J Nat Prod* 49 (4): 650–656.

Patt, P. and Wrinkler, W. (1960). The active therapeutic principle of horse chestnut (Aesculus hippocastanum). Part 2. On the chemistry of the active substance. *Arzneim Forsch/Drug Res* 10: 273-5.

Schulz V, *et al.* (1998). Rational Phytotherapy: A Physician's Guide to Herbal Medicine, 3rd ed. Berlin (Germany): Springer.

Shah, D., S. Bommer, F.H. Degenring. (1997). Aesculaforce in chronic venous insufficiency. *Schweiz Zschr Ganzheits Medizin* 9 (2): 86–91.

Sirtori, C.R. (2001). Aescin: pharmacology, pharmacokinetics and therapeutic profile. *Pharmacology Research* 44 (3):183-193.

Schulz V, *et al.,* (1998). Rational Phytotherapy: A Physician's Guide to Herbal Medicine, 3rd ed. Berlin (Germany): Springer

Vayssairat, M. *et al.,* (1996). Horse-chestnut seed extract for chronic venous insufficiency. In: *Lancet* 347 (9009): 182-183.

Yoshikawa, M. *et al.* (1996). Bioactive saponins and glycosides. III. Horse chestnut. (1): The structures, inhibitory effects on ethanol absorption and hypoglycemic activity of escins Ia, Ib, IIa, IIb, and IIIa from the seeds of *Aesculus hippocastanum* L. *Chem Pharm Bull (Tokyo)* 44 (8): 1454–1464.

Yoshikawa, M. *et al.* (1998). Bioactive saponins and glycosides. XII. Horse chestnut. (2): Structures of escins IIIb, IV, V and VI and isoescins Ia, Ib, and V, acylated polyhydroxyoleanene triterpene oligoglycosides, from the seeds of horse chestnut tree (*Aesculus hippocastanum* L., Hippocastanaceae). *Chem Pharm Bull (Tokyo)* 46(11):1764–1769.

JAMBU (JAMBUL)

Botanical Name: *Eugenia jambolana* Lam. syn. *Syzygium cumini* (Linn.) Skeels.

Natural Order: Myrtaceae.

Active Ingredients: Glucoside: jambulin, essential oil and tannin.

Non-Clinical Studies: Hypoglycaemic (Prince Menon and Pari 1998).

BIBLIOGRAPHY

Prince, P.S.; Menon, V.P. and Pari, L. (1998). Hypoglycemic activity of *Syzigium cumini* seeds: effect on lipid peroxidation in alloxan diabetic rats. *J Ethnopharmacol* 61 (1):1-7.

JATAMANSI (INDIAN SPIKENARD)

Botanical Name: *Nardostachys jatamansi* DC.

Natural Order: Valerianceae.

Active Ingredients: Volatile oil, sesquiterpene lactone: jatamansone or valeranonc, jatamols A and B (Bagchi *et al* 1991; Chatterjee *et al* (2005). iridoid glucoside: nardostachin (Ju *et al* 2003), alkaloid: actinidin, lignans (Bagchi *et al* 1991), neolignans (Bagchi *et al* 1991) and pyranocoumarin (Chatterjee *et al* 2005).

Non-Clinical Studies: Antiarrthymic, hepatoprotective (Prince, Menon and Pari 1998), hypolipidemic (Dixit *et al* 1988). anti-ulcerogenic (Rucker *et al* 1978), antihypertensive (Rucker *et al* 1978), sedative (Rucker *et al* 1978), antioxidant (Salim *et al* 2003), anti-ischemic (Salim *et al* 2003) and anti-inflammatory (Ju *et al* 2003),

Toxicity: The LD50 of jatamansone or valeranone in mice and rats: greater than 3160 mg/kg body weight (Rucker *et al* 1978)

BIBLIOGRAPHY

Ali, S. *et al.* (2000). *Nardostachys jatamansi* protects against liver damage induced by thioacetamide in rats. J *Ethnopharmacol.* 71 (3): 359-63.

Bagchi, A. *et al.* (1991). Jatamols A and B, sesquiterpenoides of *Nardostachys jatamansi* roots. *Planta Med* 57: 282.

Bagchi, A. *et al.* (1991). Neolignans and lignans of *Nardostachys jatamansi* roots, *Planta Medica* 57: 96.

Chatterjee, A. *et al.* (2005). Studies on the chemical constituents of *Nardostachys jatamansi* DC (Valerianaceae). *Indian Journal of Chemistry* 44 B (2).

Dixit, V.P. *et al.* (1988). Hypolipidaemic effects of *Curcuma longa* L and *Nardostachys jatamansi*, DC in triton-induced hyperlipidaemic rats. *Indian J Physiol Pharmacol* 32 (4): 299-304.

Ju, H.K. *et al.* (2003). Inhibitory effects of nardostachin on nitric oxide, prostaglandin E2, and tumor necrosis factor-alpha production in lipopolysaccharide activated macrophages. *Biol Pharm Bull* 26 10): 1375-8.

Rucker, G. *et al.* (1978). Isolation and pharmacodynamic activity of the sesquiterpene valeranone from *Nardostachys jatamansi* DC. *Arzneimittelforschung* 28 (1): 7-13.

Salim, S *et al.*, (2003). Protective effect of *Nardostachys jatamansi* in rat cerebral ischemia. *Pharmacol Biochem Behavor* 74 (2): 481-6.

KANTKARI (YELLOW-BERRIED NIGHTSHADE)

Botanical Name: *Solanum xanthocarpum* Schrad et Wendl. Syn *Solanum surattense* Burm.f.

Natural Order: Solanaceae.

Active Ingredients: Alkaloid: solasonine and steroid saponin: disogenin.

Non-Clinical Studies: Anti-asthmatic (Govindan *et al* 1999), anti-nociceptiven (Rahman *et al* 2003), antifertility (Dixit 1986), hypoglycemic (Gupta *et al* 2005) and spermicidal (Rao 1988)

BIBLIOGRAPHY

Govindan, S. *et al.* (1999). A pilot study on the clinical efficacy of *Solanum xanthocarpum* and *Solanum trilobatum* in bronchial asthma. *J Ethnopharmacol* 66 (2): 205-210.

Rahman, M.T. *et al.* (2003). Antinociceptive activity of the aerial parts of *Solanum xanthocarpum*. *Fitoterapia* 74 (1-2): 119-121.

Gupta, S. *et al.* (2005). Evaluation of hypoglycemic potential of *Solanum xanthocarpum* fruits in normal and streptozocin induced diabetic rats. *Eu Bull Drug Res.* 13: 51-55.

Dixit, V.P. (1986). Antifertility effects of solasodine (C27 H 43 0 2 N) obtained from *Solanum xanthocarpum* berries in male rats and dogs. *Journal of Steroid Biochemistry* 25: 27.

Rao, M.V. (1988). Effects of alcoholic extract of *Solanum xanthocarpum* seeds in adult male rats. *Indian Journal of Experimental Biology* 26 (2): 95-98.

KAPIKACHCHU

Botanical Name: *Mucuna pruriens* (L.) DC.

Natural Order: Fabaceae

Active Ingredients: 3, 4-dihydroxy-phenylalanine or L-DOPA (Pant and Joshi 1970; Brain 1976; Cargile and Doorenbos 1976; Dako and Hill 1977; Damodaran and Ramaswamy 1977; Chen 1993), alkaloids; mucuadine, mucuadinine, mucucuadinine, pruriendine, mucunine, mucunadine, nicotine (Majumdar and Zalani 1953; Ghosal, Singh and Bhattacharya 1971; Saksena and Dixit 1987), mucunin (Shelley and Arthur 1955), lecithin and 5-hydroxy-tryptamine or serotonin (Pant and Joshi 1970).

Non-Clinical Studies: Antidiabetic (Akthar *et al* 1990). aphrodisiac (Kumar *et al* 1994), analgesic and antipyretic (Iauk *et al* 1993), hypoglycemic and hypolipidemic (Pant *et al* 1968) and anti-Parkinson's (Hussain and Manyam 1997).

Clinical Studies

Parkinson's disease

A. In a study efficacy and tolerability of *Mucuna pruriens* (HP-200) was investigated in sixty patients (forty-six males and fourteen females) of Parkinson's disease. Twenty six patients were adminstred synthetic levopoa and thirty-four with HP-200. The Unified Parkinson's Disease Rating Scale was used as testing tool. Based on the findings it was concluded that HP-200 was more effective in treatment of Parkinson's disease (Anonymous 1995).

B. In a randomised, controlled, double blind crossover trial, efficiacy and tolerability of *Mucuna pruriens* was investigated in eight pateints suffering from disease. Patients were adminstred with single doses of 200/50 mg LD/CD and 15 and 30 g of mucuna preparation in randomised order at weekly intervals. L-Dopa pharma-cokinetics was determined and Unified Parkinson's Disease Rating Scale and tapping speed were obtained at baseline and repeatedly during the 4 h following drug ingestion. Dyskinesias were assessed using modified AIMS and Goetz scales. Compared with standard LD/CD, the 30 g mucuna preparation led to a considerably faster onset of effect reflected in shorter latencies to peak L-dopa plasma concentrations. It was concluded that natural source of L-dopa might possess advantages over conventional L-dopa preparations in the long term management of Parkinson's disease (Katzenschlager *et al* 2004).

Toxicity: Acetone extract of the root is toxic to insects. The drug is known to cause contact dermatitis due to presence of mucunin.

BIBLIOGRAPHY

Akthar, M.S. *et al.* (1990). Antidiabetic evaluation of *Mucuna pruriens* Linn. seeds, *J Pak Med Assoc* 40 (7): 147-150.

Kumar, A. *et al.* (1994). Aphrodisiac activity of the seeds of *Mucuna pruriens* in sexually normal rats. *Fitoterapia* 67 (1): 321-327.

Anonymous (1995). An alternative medicine treatment for Parkinson's disease: results of a multicenter clinical trial. HP-200 in Parkinson's Disease Study Group. *J Altern Complement Med* 4 68.

Brain, K.R. (1976). Accumulation of L-DOPA in cultures from *Mucuna pruriens, Plant Sci Lett* 7 (3): 157-161.

Cargile, N. and Doorenbos, N.J. (1976). New analytical procedures for L-DOPA in plant materials. *J Miss Acad Sci* 231 50-53.

Chen, Y. (1993). Determination of levodopa in *Mucuna pruriens* and *Mucuna carcarpa* by thin layer chromatography *J Chinese Herb Med* 24: 294-295.

Dako, D.Y. and Hill, D.C. (1977). Chemical and biological evaluation of *Mucuna pruriens* beans. *Nutr Rep Int* 15 (2): 239-244.

Damodaran, M. and Ramaswamy, R. (1977). Isolation of 1-3,4 dihydroxyphenylalanine from seeds of *Mucuna pruriens, Biochem J* 31: 2149.

Ghosal, S., Singh, S. and Bhattacharya, S.K. (1971). Alkaloids of *Mucuna pruriens* Chemsitry and Pharmacology. *Planta Medica* 19 (3): 279-284.

Hussain, G. and Manyam, B.V. (1997). *Mucuna pruriens* proves more effective than L-DOPA in Parkinson's disease animal disease model. *Phytotehrapy Research* 11 (6): 419-423.

Iauk, L. *et al.* (1993). Analgesic and antipyretic effects of *Mucuna pruriens. International Journal of Pharmacology* 31 (3): 213-216.

Katzenschlager, R. *et al.* (2004). *Mucuna pruriens* in Parkinson's disease: a double blind clinical and pharmacological study. *Journal of Neurology Neurosurgery and Psychiatry* 75: 1672-1677

Majumdar, D.N. and Zalani, C.D. (1953). *Mucuna pruriens* DC.: Alkaloidal Constituents. Part-3. Isolation of water soluble alkaloids and a study of their chemical and physiological characterization. *Indian J Pharm* 5. 62.

Pant, M.C. *et al.* (1968). Blood sugar and total cholesterol lowering effect of *Glycine max, Mucuna pruriens* and *Dolichos biflorus* seed diet in normal fasting albino rats. *Indian J Med Res* 56 (12): 1808-1812.

Pant, M.C. and Joshi, L.D. (1970). Identification of pharmacologically active substances in seed of *Mucuna pruriens*. DC. *Indian J Pharmacol* 2: 24.

Saksena, S. and Dixit, V.K. (1987). Role of the total alkaloids of *Mucuna prurita* Baker in spermatogenesis in albino rats. *Indian J Natl Prod* 3 (1): 3.

Shelley, W.B. and Arthur, R.P. (1956). Mucunin, the active pruritogenic proteinase of cowhage. *Biol Abstr* 30 (3): 847.

KARANJA (INDIAN BEECH)

Botanical Name: *Pongamia glabra* Vent. syn *P. pinnata* Pierre.

Natural Order: Fabaceae.

Active Ingredients: Pongamia oil: karanjin and pongamol, kanugin and demethoxykanugin.

Non-Clinical Studies: Anti-ulcerogenic (Singh *et al* 1997).

BIBLIOGRAPHY

Singh, R.K. *et al.* (1997). Pharmacological actions of *Pongamia pinnata* roots in albino rats. *Indian J Exp Biol* 35: 831-6.

KIRATIKTA (CHIRETTA)

Botanical Name: *Swertia chirayita* (Roxb ex Flem) Karst

Natural Order: Gentianaceae.

Active Ingredients: Alkaloid: gentianine, ophelic acid, bitter glucosides: amarogentin, gentopicrin, xanthones: swerchirin and mangiferin, secoiridoid glucoside: swertiamarin.

Non-Clinical Studies: Central nervous system depressant (Bhattacharya *et al* 1976). hypoglycemic (Sekar *et al* 1987; Chandrasekar, Bajpai and Mukherjee 1990; Saxena, Murthy and Mukherjee 1996) and antileshmonial (Ray *et al* 1996).

BIBLIOGRAPHY

Bhattacharya, S.K. *et al.* (1976). Chemical constituents of Gentianaceae XIX: CNS-depressant effects of swertiamarin. *J Pharm Sci* 65 (10): 1547-9.

Chandrasekar, B., Bajpai, M.B. and Mukherjee, S.K. (1990). Hypoglycemic activity of *Swertia chirayita* (Roxb ex Flem) Karst. *Indian J Exp Biol* 28 (7): 616-618.

Ray, S. *et al.,* (1996). Amarogentin, a naturally occurring secoiridoid glycoside and a newly recognized inhibitor of topoisomerase 1 from *Leishmania donovani*. *J Nat Prod* 59: 27-29.

Saxena, A.M.; Murthy, P.S. and Mukherjee, S.K. (1996). Mode of action of three structurally different hypoglycemic agents: a comparative study. *Indian J Exp Biol* 34: 406-409.

Sekar, B.C. *et al.* (1987). Effect of different fractions of *Swertia chirayita* on the blood sugar level of albino rats. *J Ethnopharmacol* 21(2):175-181.

KUSTHA (COSTUS)

Botanical Name: *Saussurea lappa* C.B. Clarke.

Natural Order: Asteraceae.

Active Ingredients: Alkaloid: saussurine (Rehm and Espig 1991), inulin, sesquiterpene lactones: cynaropicrin (Chao *et al* 2000), saussureamines A, B, C, D and E (Yoshikawa *et al* 1993), lappadilactone, costunolide, dehyrdocostuslactone, resin, aromatic oil, sterols, kushtin and costuic acid.

Non-Clinical Studies: Antibacterial (Dutta 1960), immuno-stimulant (Kulkarni and Deasi 2001), muscle relaxant (Chopra 1982), anti-inflammatory (Chao *et al* 2000; Gokhale *et al* 2002; Damre Damre and Saraf 2003), cytotoxic (Gupta and Ghatak 1967), antiulcer (Yoshikawa *et al* 1993) and anti-arthritic (Gokhale *et al* 2002).

BIBLIOGRAPHY

Chao, J.Y. *et al.* (2000). *In vitro* anti-inflammatory effects of Cynaropicrin, a sesquiterpene lactone from *Saussurea lappa*. *Eu J Pharmacol* 398: 399-407.

Damre, A.A.; Damre A.S. and Saraf, M.N. (2003). Evaluation of sesquiterpene lactone fraction of *Saussurea lappa* on transudative, exudative and proliferative phases of inflammation. *Phytother Res* 17 (7): 722-5.

Gupta, O.P. and Ghatak, B.J. (1967). Pharmacological investigations on *Saussurea lappa* (Clarke). *Indian J Med Res* 55: 1078-83.

Kulkarni, S. and Deasi, S. (2001). Immunostimulant activity of Inulin isolated from *Saussurea lappa* roots. *Indian Journal of Pharmaceutical Sciences* 63: 292-4.

Yoshikawa, M. *et al.*, (1993). Saussureamines A, B, C, D and E, new antiulcer principles from *Chinese saussurea* Radix. *Chem Pharm Bull* 41: 214-6.

Gokhale, A.B. *et al.,* (2002). Preliminary evaluation of anti-inflammatory and anti-arthritic activity of *S. lappa, A. speciosa* and *A. aspera. Phytomedicine* 9 (5): 433-437.

LATAKARANJA (FEVER NUT)

Botanical Name: *Caesalpinia bonducella* (L.) Flem syn *Caesalpinia crista* Linn.

Natural Order: Fabaceae.

Active Ingredients: Bitter principle: bonducin.

Non-Clinical Studies: Hypoglycemic (Biswas *et al* 2005).

BIBLIOGRAPHY

Biswas *et al.* (2005). Antidiabetic activity of *C. bonducella* in chronic type 2 diabetic model in Long- Evan rats and evaluation of insulin secretagogul preoperty of its fractions on isolated islets. *J Ethnopharmacol* 97(1): 117-122.

LEMON BALM

Botanical Name: *Melissa officinalis* Linn.

Natural Order: Lamiaceae.

Active Ingredients: Essential oil, flavonoids (Patora and Klimek 2002) and tritepene acids.

Non-Clinical Studies: Antimicrobial (Mimic-Ducick *et al* 2004), antioxidant (Mimic-Ducick *et al* 2004), anti-ulcerogenic (Khayyal *et al* 2001), antiviral (Cohen *et al* 1964), immuno-modulator (Drodz and Anuszewska 2003).

Clinical Studies

Amnesia

In a randomized, placebo-controlled, double-blind, balanced crossover study, dried leaf of *Melissa officinalis* resulted in modulation of mood and cognitive performance. 20 healthy, young participants received single doses of 600, 1000, and 1600 mg of encapsulated dried leaf, or a matching placebo, at 7-day intervals. The authors concluded that *Melissa officinalis* has nicotinic and muscarinic receptor-binding activity. However it lacked cholinesterase inhibitory activity (Kennedy *et al* 2003).

Herpes labialis

In a double-blind, placebo-controlled, randomized trial conducted on sixty patients having history of herpes labialis demonstated efficacy of topical use of *Melissa officinalis* (Koytchev *et al* 1999).

BIBLIOGRAPHY

Cohen, R.A. *et al.* (1964). Antiviral activity of Melissa officinalis extract. *Proc Soc Exp Biol Med* 117: 431-434.

Drodz, A. and Anuszewska, E. (2003). The effect of the *Melissa officinalis* extract on immune response in mice. *Acta Pol Pharma.* 60(6):467-70.

Kennedy, D.O. *et al.* (2003). Modulation of mood and cognitive performance following acute administration of single doses of *Melissa officinalis* (Lemon balm) with human CNS nicotinic and muscarinic receptor-binding properties. *Neuropyschopharmacol* 28 (10): 1871-81.

Khayyal, M.T. *et al.* (2001). Antiulcerogenic effect of some gastrointestinally acting plant extracts and their combination. *Arzneimittel forschung* 51 (7): 545-53.

Koytchev, R. *et al.,* (1999). Balm mint extract (Lo-701) for topical treatment of recurring herpes labialis. *Phytomed* 6 (4): 225-30.

Mimic-Ducick, N. *et al.,* (2004). Antimicrobial and antioxidant activities of *Melissa officinalis* L. (Lamiaceae) essential oil. *J Agric Food Chem* 52 (9): 2485-9.

Patora, J. and Klimek, B. (2002). Flavonoids from lemon balm (*Melissa officinalis* L., Lamiaceae). *Acta Pol Pharma* 59 (2): 139-43.

LILY OF THE VALLEY

Botanical Name: Convallaria majalis L.

Natural Order: Liliaceae.

Active Ingredients: Cardiac glycosides: convallotoxin (Chernov 1960) and convallerin, convalloside (Wagner and Wiesenauer 1995). convallamaroside (Wagner and Wiesenauer 1995), convallotoxol (Kreen *et al* 1992) flavonoid glycosides.

Non-Clinical Studies: Cardiac insufficiency (Chernov 1960; Frantsuzowa *et al* 1980).

Toxicity: LD_{50} of convallotoxin: 0.08mg/kg (Wagner and Wiesenauer, 1995).

BIBLIOGRAPHY

Anon (2000). PDR for Herbal Medicines. Medical Economics Company Inc, Montvale, NJ: 812-813.

Chernov, V.I. (1960). Therapy of cardiac insufficiency with convallotoxin. *Ter Arkh* 32: 26-34.

Frantsuzowa, S.B. *et al.,* (1980). Action of convallotoxin on cardiac hemodynamics in experimental stenosis of the pulmonary trunk. *Farmakol Toksikol* 43 (6): 674-9.

Kreen, L. *et al.* (1992). HPLC, separation and quantitative determination of cardenolides. In Herba Convallariae. *Planta Med* 58 (7) A82.

Wagner, H. and Wiesenauer, M. (1995). *Phytotherapy*, Phytopharma Homeopathy, Fischer –Verlag, Stuttgart, Jena, New York.

MAD DOG SKULLCAP

Botanical Name: Scutellaria lateriflora L.

Natural Order: Lamiaceae.

Active Ingredients: Flavonoids: baicalin (Chang, Chen and Lu, 2002), baicalein (Chang Chen and Lu 2002), hispulidin, scutellarin, wogonin (Chang Chen and Lu 2002).

Non-Clinical Studies: Anxiolytic (Awad *et al* 2003). antihistaminic, anti-arthritic, anti-inflammatory (Kimura *et al* 1985), anticancer (Chang, Chen and Lu 2002; Suk *et al* 2003).

Clinical Studies

Anxiety neurosis

A double blind, placebo-controlled study of healthy subjects demonstrated anxiolytic activity of *Scutellaria lateriflora* (Wolfson and Hoffmann 2003).

BIBLIOGRAPHY

Kimura, Y., *et al.* (1985). Studies on scutellariae radix; XII. Effects of various flavonoid on arachidonate metabolism in leukocytes. *Plant Medica* 2. 132-136.

Kurnakov, B.A. (1957). Pharmacology of skullcap, *Farmakologia I Tosikologia* 20 (6): 79-80.

Wolfson, P. and Hoffmann, D.L. (2003), An investigation into the efficacy of *Scutellaria lateriflora* in healthy volunteers. *Altern Ther Health Med* 9 (2): 74-78.

Awad, R., *et al.* (2003) Phytochemical and biological analysis of skullcap (*Scutellaria lateriflora* L.): a medicinal plant with anxiolytic properties. *Phytomedicine* 10 (8): 640-649.

Takeshita, K. *et al.* (2001). Pneumonitis induced by ou-gon (scullcap). *Intern Med* 40 (8): 764-768.

Suk, K. *et al.* (2003). Flavonoid baicalein attenuates activation-induced cell death of brain microglia. J *Pharmacol Exp Ther* 305 (2): 638-645.

Chang, W.H.; Chen, C.H. and Lu, FJ. (2002). Different effects of baicalein, baicalin and wogonin on mitochondrial function, glutathione content and cell cycle progression in human hepatoma cell lines. *Planta Med* 68 (2): 128-132.

MANDOOKPARNI

Botanical Name: *Centella asiatica* (L.) Urban Syn. *Hydrocotyle asiatica* Linn.

Natural Order: Apiaceae

Active Ingredients: Tritrepenes: asiaticoside, asiaticoside A, asiaticoside B brahmoside, brahminoside, centelloside, indocentelloside, medacassoside, thankuniside (Datta and Basu 1962; Rastogi and Dhar 1963; Datta and Basu 1967; Boiteau *et al* 1967; Luo and Jin 1981; Askawa *et al* 1982; Prum *et al* 1983; Chaudhary *et al* 1987; Krishan *et al* 1994; Bonte *et al* 1994). triterpene acids: asiatic acid, centic acid, centellic acid, centoic acid, indocentoic acid, thankunic acid, medecassic acid (Datta and Basu 1962; Boiteau *et al* 1967; Datta and Basu 1967; Singh and Rastogi 1969; Bonte *et al* 1994). Triterpene glycosides; triterpene oligoglycosides: centellasaponins B, C and D (Mastuda *et al* 2001), alkaloid: hydrocotyline, flavonoids, pectin and volatile oil.

Non-Clinical Studies: Wound-healing (Suguna *et al* 1966; Marastoni *et al* 1982; Del Vecchio *et al* 1984; Brinkhaus B *et al.,*1998; Shukla *et al* 1999). antioxidant (Shukla *et al* 1999), gastro protective (Brinkhaus *et al* 1998) varicose veins (Arpaia *et al* 1990; Cesarone *et al* 1994; Monteverde *et al* 1987; Pointel *et al* 1987), anti-cancer (Babu and Padikkala 1993), antifilarial (Chakraborty, Sinha, Babu and Sukul 1996), antianxiety (Diwan Karawande and Singh 1991). anti-ulcer (Sharma *et al* 1995), anti-stress (Sharma *et al* 1995); antifeedent (Srivastava, Shukla and Tripathi 1997) and fertility enhancer (Dutta *et al* 1968).

Clinical Studies

1. Venous insufficiency

A. A randomised, placebo-controlled, double-blind multi-centre trial was done to investigated the efficiacy of a *Centella asiatica* extract (60 mg or 120 mg/day for 2 months) in 94 patients diagnosed with venous insufficiency. Significant and dose-dependant improvements in subjective and plethysmographic (extremity volume) parameters were seen with both dosages of *Centella asiatica* treatment (Pointel *et al* 1987).

B. A double-blind study with 40 patients suffering chronic venous insufficiency found that *Centella asiatica* extract (60 mg/day for 30 days) significantly improved ankle circumference, vascular tone and leg volume compared with baseline (Monteverde *et al* 1987).

2. Varicose veins

A study was done to study the efficiacy of *Centella asiatica* in varicose veins. The topical application of a *Centella asiatica* extract 3 times daily was found to be effective in improving both objective and subjective symptoms including vascular tone assessed by ultrasound. This double-blind placebo-controlled study involved 80 patients with various venous disorders, in particular varicose veins and post-thrombotic syndrome (Allegra *et al* 1981).

3. Venous hypertension

A. In a trial, 62 patients diagnosed with venous hypertension were recommded *Centella asiatica* extract (90 mg or 180 mg daily) or placebo. A dose-dependant reduction in capillary filtration rate and ankle circumference was seen at the end of the 4-week study period, and changes were significant compared with placebo (Belcaro *et al* 1990).

B. Another study involving 87 patients with chronic venous hypertensive microangiopathy found that microcirculatory parameters were significantly improved compared with placebo in patients taking *Centella asiatica* extract (60 mg or 120 mg daily) after 60 days (Cesarone *et al* 1994).

4. Stretch marks

A double-blind study involving 80 pregnant women investigated the ability of a cream to prevent striae gravidarum. The cream, which contained a *Centella asiatica* extract as well as tocopherol and collagen-elastin hydrolysate, was applied once daily for 12 weeks. Women using the cream developed significantly fewer stretch marks than women using placebo (Mallol *et al* 1991).

5. Wound-healing

Several clinical studies investigating role of *Centella asiatica* in the treatment of skin wounds found increased wound perfusion and proliferation of granulation tissue and accelerated wound closure in the case of skin grafts (Brinkhaus *et al* 1998).

6. Cellulite

A clinical study reported that the topical application of a *Centella asiatica* extract 4 times daily resulted in significant improvements in 85% of 60 patients with cellulite (Pereira 1979).

7. Mental ability

A study reported favourable effect of *Centella asiatica* on the general mental ability of mentally retarded children (Rao, Srinivasan and Rao 1973).

Pharmacokinetics

Twelve healthy volunteers received each treatment following a randomized cross-over design with trials separated by a 3-week interval. The time of peak plasma concentration was not affected by dosage difference or by treatment scheme. Differences in peak plasma concentration and area under the concentration vs. time curve from 0 to 24 h (AUC0-24) calculated after 30 or 60 mg administration (single dose) were accounted for by the different dose regimen. However, after chronic treatment with both 30 and 60 mg, peak plasma concentrations, AUC0-24 and half-life were significantly higher than those observed after the corresponding single dose administration. This phenomenon could be explained by a metabolic interaction between asiatic acid and asiaticoside, which is transformed into asiatic acid *in vivo* (Grimaldi *et al* (1990).

Adverse drug reaction: Contact dermatitis (Eun *et al* 1985; Bilbao *et al* 1995; Delucia *et al* 1997).

BIBLIOGRAPHY

Allegra, C. *et al.* (1981). *Centella asiatica* extract in venous disorders of the lower limbs. Comparative clinico-instrumental studies with a placebo. *Clinica Terapeutica* 99: 507-13.

Appa Rao, M.V.R.; Srinivasan, K. and Rao, K.T. (1973). The effect of Mandookparni (*Centella asiatica*) on the general mental ability (Medhya) of mentally retarded children, *J Res Indian Med* 8 (4): 9-16.

Arpaia, M.R. *et al.* (1990). Effects of *Centella asiatica* extract on muco-polysaccharide metabolism in subjects with varicose veins, *Int. J Clin Pharmacol Res* 10 (4): 229-233.

Askawa, Y. *et al.* (1982). Mono and sesquiterpenoids from Hydrocotyle and Centella species, *Phytochemistry* 21 (10): 2590-2592.

Babu, T.D. and Padikkala, J. (1993). Anticancer activity of an active principle form *Centella asiatica. Amla Research Bulletin* 13 46-49.

Belcaro, G.V. *et al.* (1990). Capillary filtration and ankle edema in patients with venous hypertension treated with TTFCA. *Angiology* 41:12-8.

Bilbao, I. *et al.* (1995). Allergic contact dermatitis from butoxyethyl nicotinic acid and *Centella asiatica* extract. *Contact Dermatitis* 33 (6):435-436.

Boiteau, P. *et al.* (1967). Isolation of a new triterpenic acid from Centella asiatica (L.) Urb. Of Madagascar: Madecassic acid, C.R. *Acad Sci Hebd Seances Sci* D 264 (2): 407-410.

Bonte, F. *et al.* (1994). Influence of asiatic acid, madecassic acid and asiaticoside on human collagen-I synthesis, *Planta Med* 60 (2): 133-135.

Brinkhaus, B. *et al.* (1998). Centella *asiatica* in traditional and modern phytomedicine - A pharmacological and clinical profile-Part II: Pharmacological and therapeutical profile, conclusions. *Perfusion* 11: 508-20.

Cesarone, M.R. *et al.* (1994). Activity of *Centella asiatica* in venous insufficiency, *Minerva cardioangiol* 40 (4) 137-143.

Cesarone MR *et al.* (1994). The microcirculatory activity of *Centella asiatica* in venous insufficiency. A double-blind study. *Minerva Cardioangiologica* 42: 299-304.

Chakraborty, T.; Sinha Babu, S.P. and Sukul, N.C. (1996). Preliminary evidence of antifilarial effect of *Centella asiatica* on Canine dirofilariasis. *Fitoterapia* 67 (2): 101-112.

Chaudhary, A.K.A. *et al.* (1987). Biological activity of the alcohol extracts and the glycosiders of *Hydrocotyle asiatica. J Bangladesh Acad Sci* 11(1) 75-82.

Darms, F. *et al.,* (1979). Use of a titrated extract6 of *Centella asiatica* in chronic hepatic disorders. *Sem Hos* 55 (37-38): 1749-1750.

Datta, T. and Basu, U.P. (1962). Triterpenoids Part-1. Thankuniside and Thankunic acid: A new triterpene glycoside and acid from *Centella asiatica* L. (Urb.). *J Sci Indudtr Res* 21B 239.

Datta, T and Basu, U.P. (1967). Terpenoids IV: Isolation and identification of Asiatic acid from *Centella asiatica* Linn. *Indian J Chem* 5 586.

Del Vecchio, A. *et al.,* (1984). Effect of *Centella asiatica* on the biosynthetic activity of fibroblasts in culture. *Farmaco* [Prat], 39 (10): 355-364.

Delucia, R. *et al.,* (1997). Pharmacological and Toxicological studies on *Centella asiatica* extract, *Fitoterapia* 68 (5): 413-416.

Diwan, P.V.; Karawande, I. and Singh, A.K. (1991). Antianxiety profile of Mandukparni *(Centella asiatica)* in animals. *Fitoterapia* 62 (3) 253-257.

Dutta, T. *et al.* (1968). Crude extracts of *Centella asiatica* and products derived from its glycosides as oral infertility agents. *Indian J Exp Biol* 6 (3): 181-182.

Eun, H.C. *et al.* (1985). Contact dermatitis due to medacassol. *Contact Dermatitis* 13 (5): 310-313.

Grimaldi, R. *et al.* (1990). Pharmacokinetics of the total triterpenic fraction of *Centella asiatica* after single and multiple adminstrations to healthy volunteers-A new assay for asiatic acid. *J Ethnopharmacol* 28 (2): 235-241.

Krishan Lal *et al.* (1994). HPLC analysis of asiaticoside in *Centella asiatica. Indian Journal of Pharmaceutical Sciences* 56 (4): 154.

Luo, S and Jin. H (1981). Isolation and identification of medacassoside in *Centella asiatica. Zhongcaoyao* 12 (6): 5-6.

Mallol, J. *et al.* (1991). Prophylaxis of striae gravidarum with a topical formulation. *International Journal of Cosmetic Science* 13:51-57.

Marastoni, F. *et al.* (1982). *Centella asiatica* extract in venous pathology of the lower limbs and its evaluation as compared to tribenoside. *Minerva Cardioangiol* 30 (4): 201-207.

Mastuda, H. *et al.* (2001). Medicinal foodstuffs. XXVII. Saponin constituents of gotu kola (2): structures of new ursane- and oleanane-type triterpene oligoglycosides, centellasaponins B, C, and D, from *Centella asiatica* cultivated in Sri Lanka. *Chem Pharm Bull* (Tokyo) 49 (10): 1368-71.

Monteverde A *et al.* (1987). Comparison between extract of *Centella asiatica* and O-(b-Hydroxyethyl) rutoside in the treatment of venous insufficiency of the lower limbs. *Acta Therapeutica* 13: 629-36.

Pereira, I.C. (1979). Trattamento da celulite pela *Centella asiatica. F Med* (BR) 5: 401-4.

Pointel, J.P. *et al.* (1987). Titrated extract of *Centella asiatica* (TECA) in the treatment of venous insufficiency of the lower limbs. *Angiology* 38: 46-50.

Prum *et al.* (1983). Flavonoid glycosides from *Centella asiatica* L. (Umbelliferae). *Pharmazie* 38 (6): 423.

Rastogi, R.P. and Dhar, M.L. (1963). Chemical examination of *Centella asiatica* Linn. Part-II. Brahmoside and Brahminoside. *Indian J Chem* 1 267.

Rastogi, R.P. and Dhar, M.L. (1963). The saponins of *Centella asiatica* Linn. *Bull Nat Inst Sci* India 23: 57.

Sharma, D.N.K. *et al.* (1995). Antiulcer activity of *Tinsopora cordifolia* Miers. and *Centella asiatica* Linn. extracts. *Phytotherapy Research* 9 (8): 589-590.

Sharma, D.N.K. *et al.* (1995). Antistress activity of *Tinsopora cordifolia* Miers. and *Centella asiatica* Linn. extracts. *Phytotherapy Research* 10 (2): 181-183.

Shukla, A. *et al.* (1999). Asiaticoside-induced elevation of antioxidant levels in healing wounds. *Phytother Res* 13 (1): 50-4.

Singh, B and Rastogi, R.P. (1969). Chemcial examination of *Centella asiatica* Linn. III., Constitution of brahmic acid. *Phytochemistry* 7: 1385.

Srivastava, R.; Shukla, Y.N. and Tripathi, A.K. (1997). Antifeedent compounds from *Centella asiatica. Fitoterapia* 68 (1): 93-94.

Suguna, L. *et al.* (1966). Efffects of *Centella asiatica* extract on dermal wound healing in rats, *Indian J Exp Biol* 34 (12): 1208-1211.

MANJISTHA

Botanical Name: *Rubia cordifolia* Linn.

Natural Order: Rubiaceae

English Name: Indian Madder

Active Ingredeints: Resin, colouring matter: manjistin, purpurin, xanthopupurin alizarin, peseudopurpurin, dihydroxy anthraquinone: rubidianin (Tripathi, Sharma and Manickam 1997), anthraquinones: cordifoliol and cordifodiol (Abdullah *et al* 2003), triterpene (Kasture, Deshmukh and Chopde 2000).

Pre-Clinical Studies: Anti-cancer (Adwankar *et al* 1980), antiplatelet (Tripathi, Pandey and Shukla 1993), antioxidant (Tripathi, Sharma and Manickam 1997; Tripathi and Sharma 1998), anxiolytic and anticonvulsant (Kasture, Deshmukh and Chopde 2000).

BIBLIOGRAPHY

Abdullah, S.T. *et al.* (2003). Two new anthraquinones from the roots of *Rubia cordifolia* Linn. *Pharmazie* 58 (3): 216-7.

Adwankar, M.K., *et al.* (1980). Anti-cancer activity of the extracts of *Rubia cordifolia* Linn. *Indian J Exp Biol* 18 (1): 102.

Kasture, V.S.; Deshmukh, VK., and Chopde, C.T. (2000). Anticonvulsant and behavioral actions of triterpene isolated from *Rubia cordifolia* Linn. *Indian J Exp Biol* 38 (7): 675-80.

Tripathi, Y.B.; Pandey, S. and Shukla, S.D. (1993). Anti-platelet activating factor property of *Rubia cordifolia* Linn. *Indian J Exp Biol* 31 (6): 533-5.

Tripathi, Y.B.; Sharma, M. and Manickam, M. (1997). Rubidianin, a new antioxidant from *Rubia cordifolia. Indian J Biochem Biophys* 34: 302-6.

Tripathi, Y.B. and Sharma, M. (1998). Comparison of the antioxidant action of the alcoholic extract of *Rubia cordifolia* with rubiadin. *Indian J Biochem Biophys* 35 (5): 313-6.

MARSH PEPPER KNOTWEED

Botanical Name: *Polygonum hydropiper* Linn.

Natural Order: Polygonaceae

Active Constituents: Glycoside: Hydropiperoside (Fukjujama *et al* 1983). flavonoids (Penq *et al* 2003) and sesquiterpenoids (Tripathi *et al* 1999).

Non-Clinical Studies: Antioxidant (Penq *et al* 2003).

Adverse Drug Reactions: Contact dermatitis (Steyn 1934).

BIBLIOGRAPHY

Fukjujama *et al.,* (1983). Hydropiperroside, a novel coumaroly glycoside from the root of *Polygonuim hydropiper*. *PH* 22: 549-552.

Penq, J.F. *et al.,* (2003). Antioxidant flavonoids from leaves of *Polygonum hydropiper* L. *Phytochem* 62 (2): 219-28.

Steyn, D.G. (1934). The Toxicology of Plants in South Africa together with a consideration of poisonous foodstuffs and fungi. South Africa: Central News Agency.

Tripathi, A.K. *et al.,* (1999). Persistency of bioactive fractions of Indian plant, *Polygonum hydropiper* as an insect feeding deterrent. *Phytotherapy Res* 13 (3): 239-241.

MEADOWSWEET

Botanical Name: *Filipendula ulmaria* (Linn.) Maxim.

Natural Order: Scrophulariaceae.

Active Ingredients: Flavonoids: spiraeoside (Hia and Lia 1980; Tits *et al* 1992; Zeylstra 1998). salicylates: salicin, salicylaldehyde, methyl salicylate (Zeylstra 1998).

Non-Clinical Studies: Antibiotic (Csedo *et al* 1993). immunomodulator (Halkes *et al* 1997).

BIBLIOGRAPHY

Halkes, *et al.* (1997). *In vitro* immunomodulatory activity of *Filipendula ulmaria. Phytother Res* 11 (7): 518-520.

Hia, H. and Lia, L. (1980). Phytochemical study of the dropworts, *Filipendula ulmaria* and *F. hexapetala*, from the flora of Lvov Province. *Farm Zh* (1): 50-2.

Tits, M. *et al.* (1992). Densitometric evaluation of spiraeoside after derivatization in flowers of *Filipendula ulmaria* (L.) Maxim. *J Pharm Biomed Anal* 10 (10-12): 1085-8.

Csedo, K. *et al.,* (1993). The antibiotic activity of *Filipendula ulmaria. Planta Med* 59 (7): A675.

Zeylstra, H. (1998). *Filipendila ulmaria. Br J Phytotherap* 5: 8–12.

MEZEREON

Botanical Name: *Daphne mezereum* Linn.

Natural Order: Thymelaceae.

Active Ingredients: Diterpenes: daphnetoxin and mezerein

Non-Clinical Studies: Anti-tumor (Borton *et al* 1989).

BIBLIOGRAPHY

Borton, K *et al.* (1989). The effects of the anti-tumor agent mezerein on the cytotoxic capacity and oxidative metabolism of human blood cells. *Invest New Drug* 7 (2-3): 179-88.

Gffney, E.V. *et al.,* (1984). Stimulation of diploid fibroblast growth with serum-free medium conditioned by mezerein-treated monocytic leukemia cells. *Journal of Leukocyte Biology* 35 (5): 489-500.

NAHI BOOTI (MAJAMAK BOOTI)

Botanical Name: *Enicostemma littorale* Blume

Natural Order: Gentiaceae

Active Ingredeints: Bitter principle (swertimarine), two alkaloids (one gentianine and other's name not confirmed) and ophelic acid and tannins.

Pre-Clinical Studies: Antidiabetic (Murali *et al* 2003) and hypoglycemic (Vijayvargia, Kumar and Gupta 2000), Maroo *et al* 2003; Maroo *et al* 2003), antitumour (Kavimani and Manisenthlkumar 2000).

Human Studies: A human study has reported efficacy of *Enicostemma littorale* in diabetes mellitus type-2 (Upadhay *et al.,* 2004). Pills prepared from *Enicostemma littorale* were administered to 84 patients with type 2 diabetes for 3 months. Estimation of various biochemical parameters showed that *E. littorale* reduced blood glucose as well as serum insulin levels and prevented the progression of complications in diabetic patients. Significant improvement in kidney function, lipid profile, and blood pressure was observed.

BIBLIOGRAPHY

Rai, J. and Thakar, K.A. (1998). Chemical Investigation of *Enicostemma littorale* Blume. *Current Science* 35. 148-149.

Kavimani, S. and Manisenthlkumar, K.T. (2000). Effect of methanolic extract of *Enicostemma littorale* on Dalton's ascitic lymphoma. *J Ethnopharmacol* 71 (1-2): 349-52.

Maroo, J. *et al.,* (2003). Antidiabetic Efficacy of *Enicostemma littorale* Methanol Extract in Alloxan-induced Diabetic Rats. *Pharmaceutical Biology* 41 (5): 388-91.

Maroo, J., Vasu, V.T. and Gupta, S. (2003). Dose dependent hypoglycemic effect of aqueous extract of *Enicostemma littorale* Blume in alloxan induced diabetic rats. *Phytomedicine* 10(2-3): 196-199.

Murali, B. *et al.* (2003). Effect of chronic treatment with *Enicostemma littorale* in non-insulin-dependent diabetic (NIDDM) rats. *J. Ethnopharmacol* 81 (2): 199-204.

Upadhay, U.M. *et al.* (2004). Efficacy of *Enicostemma littorale* in type 2 diabetic patients. *Phytotherapy research* 18 (3): 233-235.

Vijayvargia, R.; Kumar, M. and Gupta, S. (2000). Hypoglycemic effect of aqueous extract of *Enicostemma littorale* Blume (chhota chirayata) on alloxan induced diabetes mellitus in rats. *Indian J Exp Biol* 38 (8): 781-784.

NIMBA (MARGOSA)

Botanical Name: *Azadirachta indica* A. Juss.

Natural Order: Meliaceae

Active Ingredients: Terpenes: azadirachtin, nimbin, nimbinin, nimbidin, bitter principle: margosin and oil: margosa oil, catechin and tannin.

Non-Clinical Studies: Antihyperglycemic (Chattaopadhyay, 1996). anti-inflammatory (Chattaopadhyay 1998), anxiolytic (Jaiswal, Bhattacharya and Acharya 1994), hypoglycemic (Chattopadhyay 1999) and hypotensive (Chattaopadhyay 1997).

BIBLIOGRAPHY

Chattaopadhyay R.R. (1998). Possible biochemical mode of anti-inflammatory action of *Azadirachta indica* A Juss in rats. *Indian J Exp Biol* 36: 418-420.

Chattaopadhyay, R.R. (1996). Possible mechanism of antihyperglycemic effect of *Azadirachta indica* leaf extract Part IV. *Gen Pharmacol* 27: 431-434.

Chattaopadhyay, R.R. (1997). Effect of *Azadirachta indica* hydroalcohlic leaf extract on the cardiovascular system. *Gen Pharmacol* 28: 449-451.

Chattopadhyay, R.R. (1999). A comparative evaluation of some blood sugar lowering agents of plant origin. *Journal of Ethnopharmacology* 67: 367-372.

Jaiswal, A.K.; Bhattacharya, S.K. and Acharya, S.B. (1994). anxiolytic activity of leaf extract of *Azadirachta indica* in rats. *Indian J Exp Biol* 32: 489-91.

NIRGUNDI (FIVE-LEAVED CHASTLE)

Botanical Name: *Vitex negundo* Linn.

Natural Order: Verbenaceae.

Active Ingredients: Volatile oil (Stylian 1998; Maria 2001), resin (Chawla *et al* 1992), triterpenoids (Chawla *et al* 1992), norditerpenoid: vitedoin B (Ono *et al* 2004), phenylnaphthalene-type lignan alkaloid: vitedoamine A (Ono *et al* 2004).

Non-Clinical Studies: Anti-inflammatory (Jain and Pande 1976; Sharma and Singh 1980; Chawla *et al* 1992; Jana Chattopadhyay and Shaw 1999; Dharmasiri *et al* 2003), analgesic (Sharma and Singh 1980; Gupta *et al* 1997; Dharmasiri *et al* 2003), antihistaminic (Sharma and Singh 1980), anti-stress (Tandon and Gupta 2005), antioxidant (Ono *et al* 2004), anticonvulsant (Gupta *et al* 1997), immunostimulant (Stylian 1998), cytotoxic (Castillo 2003).

BIBLIOGRAPHY

Castillo, F.D. (2003). Thesis seminar: "Cytotoxic constituents of *Vitex negundo* and *Picramnia latifolia*". MDCH 595/PMPG 595 Seminars in Medicinal Chemistry/ Pharmacognosy.

Chawla, A.S. *et al.*, (1992). Chemical investigation and anti-inflammatory activity of *Vitex negundo* seeds. *J Nat Prod* 55 (2): 163-7.

Dharmasiri, M.G. *et al.*, (2003). Anti-inflammatory and analgesic activities of mature fresh leaves of *Vitex negundo*. *J Ethnopharmacol* 87 (2-3): 199-206.

Gupta, M. *et al.*, (1997). CNS activity of petroleum ether extract of vitex negundo linn in mice. *Indian Journal of Pharmaceutical Sciences* 59 (5): 240-5

Jain P.K. and T.N. Pande. (1976). *J Res Ind Med Yoga Homeo.* 11: 2.

Jana, U.; Chattopadhyay, R.N. and Shaw, B.P. (1999). Preliminary studies on anti-inflammatory activity of *Vitex negundo* Linn in albino rats. *Indian J Pharmacol* 31 232-233.

Kaushik, R.D.; Kumar A. and Arora, C. (2003). 'Fractionation, Characterization and Evaluation of biocidal potential of active principles of leaves of *Vitex negundo'*, *Asian Journal of Chemistry* 15 (3 & 4): 1659-1664.

Maria, G. (2001). Application of Thermal Desorption Gas Chromatography for the Analysis of the Volatile Organic Compounds in Fresh and Dried Leaves of *Vitex negundo* and *Psidium guajava.* (Undergraduate thesis).

Ono, M. *et al.,* (2004). Lignan derivatives and a norditerpene from the seeds of *Vitex negundo. J Nat Prod* 67 (12): 2073-5.

Sharma, A.K. and R.H. Singh (1980). *Bull Med Ethnobot Res.* 1: 262.

Stylian, G. (1998). *Vitex negundo* Monograph. *Australian Journal of Medical Herbalism.*

Tandon,V. and Gupta, R.K. (2005). Effect of *Vitex negundo* on oxidative stress. *Indian Journal of Pharmacology* 37 (1): 38-40.

OLIVE

Botanical Name: *Olea europaea* Linn.

Natural Order: Oleaceae.

Active Ingredients: Irodoid glycosides: oleuropein and hydroxytyrosol (Fleming *et al* 1973; Bianco *et al* 1992; Bisignano *et al* 1999; Edgecombe *et al* 2000; Ranalli *et al* 2005), flavonoids (Nychas, Tassou and Board 1990), triterpenes, monoterpene: elonic acid (Renis 1969; Elliott Buthala and DeYoung 1969).

Non-Clinical Studies: Antimicrobial (Fleming *et al* 1973; Rodriguez *et al* 1988; Nychas, Tassou and Board 1990; Bisignano *et al* 1999), antiviral (Renis 1969) and antioxidant (Edgecombe *et al* 2000).

BIBLIOGRAPHY

Bianco, A. *et al.* (1992). Partial synthesis of oleuropein. *J Nat Prod* 55 (6): 760-766.

Bisignano, G. *et al.* (1999). On the *in-vitro* antimicrobial activity of oleuropein and hydroxytyrosol. *J Pharm Pharmacol* 51 (8): 971-4.

Edgecombe, S.C. *et al.* (2000). Oleuropein, an Antioxidant Polyphenol from Olive Oil, Is Poorly Absorbed from Isolated Perfused Rat Intestine. *Journal of Nutrition* 130: 2996-3002

Elliott, G.A.; Buthala, D.A. and DeYoung, E.N. (1969). Preliminary Safety Studies with Calcium Elenolate, an Antiviral Agent. *Antimicrobial Agents and Chemotherapy* 173-76.

Fleming, H.P. *et al.* (1973). Antimicrobial Properties of Oleuropein and Products of Its Hydrolysis from Green Olives. *Appl Microbiol* 26 (5): 777–782.

Nychas, G.J.E., Tassou, S.C. and Board, R.G. (1990). Phenolic extract from olives: inhibition of *Staphylococcus aureus*. *Letters in Applied Microbiology* 10: 217-220.

Ranalli, A. *et al.* (2005). Factors affecting the contents of Iridoid Oleuropein in Olive Leaves (*Olea europaea* L.) *J Agric Food Chem* S0021-8561 (05) 01647-X.

Renis, H.E. (1969). *In vitro* Antiviral Activity of Calcium Elenolate. *Antimicrobial Agents and Chemotherapy*. 167-72.

Rodriguez, M.M. *et al.* (1988). Effect of extracts obtained from olive oil mill waste waters on *Bacillus megaterium* ATCC 33085. *Journal of Applied Bacteriology* 64: 219-26.

PALASHA
(FLAME OF THE FOREST)

Botanical Name: *Butea monosperma* (Lam.) Taub. Syn. *Butea frondosa* Roxb.

Natural Order: Fabaceae.

Active Ingredients: Gallic acid, kino tannic acid, sesquiterpene lactone: palasonin (Kumar *et al* 1995; Fietz *et al* 2002). glucosides: butrin, butein, corcopsin, isocoreopsin, sulphurein, monospermoside and isomonospermoside.

Non-Clinical Studies: Anthelmintic (Kumar *et al* 1995). anti-stress (Bhatwadekar *et al* 1999) and anticonvulsant (Kasture *et al* 2002).

BIBLIOGRAPHY

Bhatwadekar, A.D. *et al.* (1999). Antistress activity of Butea monosperma flowers. *Indian Journal of Pharmacology* 31(2): 153-155.

Kasture, V.S. *et al.* (2002). Anticonvulsive activity of *Butea monosperma* flowers in laboratory animals. *Pharmacol Biochem Behavor* 72(4):965-72.

Kumar, D. *et al.* (1995). Possible mechanism of anthelmintic action of palasonin on *Ascaridia galli*. *Indian J Pharmacol* 27: 161-8. Short Communication.

Fietz, O. *et al.,* (2002). (R)-(+)-palasonin, a cantharidin-related plant toxin, also occurs in insect hemolymph and tissues. *Journal of Chemical Ecology* 28 (7): 1315-27.

PASSION FLOWER

Botanical Name: *Passiflora incarnata* Linn.

Natural Order: Passifloraceae.

Active Ingredients: Flavonoids: apigenin, chyrsin, homoorientin, isovitexin, kaempferol, luteolin, orientin, quercitin, rutin, saponaretin, saponarin, vitexin and various glycosides (Neu 1954; Menghini *et al* 1993), alkaloid: passiflorine (Neu 1954). 3-methyl-4-carbolin (2'-methyl-(pyridino-3', 4': 2, 3-indole)) (Neu 1954), harmane, harmine (Bennati 1971), novel tri-substituted benzoflavone moiety (Dhawan *et al* 2003), maltol (Kimura and Murata 1974).

Non-Clinical Studies: Anti-inflammatory, anti-ageing (Dhawan, Kumar and Sharma 2002), anti-tussive (Dhawan and Sharma, 2002), anxiolytic (Dhawan, Kumar and Sharma 2001), antispasmodic (Speroni and Minghetti 1988), motility-inhibiting, central nervous system depressant (Dhawan *et al* 2004), anti-asthmatic (Dhawan, Kumar and Sharma 2003), aphrodisiac (Dhawan, Kumar and Sharma 2003) and sedative (Soulamin *et al* 1997).

Clinical Studies

Anxiety

The study was performed on 36 out-patients diagnosed with Details GAD using DSM IV criteria. Patients were allocated in a random fashion: 18 to the *Passiflora extract* 45 drops/day plus placebo tablet group, and 18 to oxazepam 30 mg/day plus placebo drops for a 4-week trial. *Passiflora extract* and oxazepam were effective in the treatment of generalized anxiety disorder. No significant difference was observed between the two protocols at the end of trial. Oxazepam showed a rapid onset of action. On the other hand, significantly more problems relating to impairment of job performance were encountered with subjects on oxazepam. The results suggest that Passiflora extract is an effective drug for the management of generalized anxiety disorder, and the low incidence of impairment of job performance with Passiflora extract compared to oxazepam is an advantage (Akhondzadeh *et al* 2001).

Opiate withdrawal

A total of 65 opiates addicts were assigned randomly to treatment with passiflora extract plus clonidine tablet or clonidine tablet plus placebo drop during a 14-day double-blind clinical trial. All patients met the DSM IV criteria for opioid dependence. The fixed daily dose was 60 drops of passiflora extract and a maximum daily dose of 0.8 mg of clonidine administered in three divided doses. The severity of the opiate withdrawal syndrome was measured on days 0, 1, 2, 3, 4, 7 and 14 using the Short Opiate Withdrawal Scale (SOWS). Both protocols were equally effective in treating the physical symptoms of withdrawal syndromes. However, the passiflora plus clonidine group showed a significant superiority over clonidine alone in the management of mental symptoms. These results suggested that passiflora extract may be an effective adjuvant agent in the management of opiate withdrawal (Kashani *et al* 2001).

Toxicity: A 34-year-old female developed severe nausea, vomiting, drowsiness, prolonged QTc, and episodes of nonsustained ventricular tachycardia following self-administration of *Passiflora incarnata*. (Fisher *et al* 2000). The extract prolonged the sleeping time induced by pentobarbital and reduced significantly the spontaneous motor activity. These effects were observed with doses of 75 and 150 mg/kg of extract. Its LD50 value was estimated as 456 mg/kg i.p. for mice (Oga *et al* 1984).

BIBLIOGRAPHY

Akhondzadeh, S. *et al.* (2001). Passionflower in the treatment of generalized anxiety: a pilot double-blind randomized controlled trial with oxazepam. *J Clin Pharm Ther* 26 (5): 363-7.

Aoyagi, N.; Kimura, R. and Murata, T. (1974). Studies on *Passiflora incarnata* dry extract. I. Isolation of maltol and pharmacological action of maltol and ethyl maltol. *Chem Pharm Bull* (Tokyo) 22 (5): 1008-13.

Bennati, E. (1971). Quantitative determination of harmane and harmine in the extract of *Passiflora incarnate. Boll Chim Farm* 110 (11): 664-9.

Cronin, J. (2003). Passionflower: Reigniting male libido and other potential uses. *Alternative and Complementary Therapies* 89-92.

Dhawan, K. and Sharma, A. (2002). Antitussive activity of the methanol extract of Passiflora incarnata leaves. *Fitoterapia* 73 (5): 397-9.

Dhawan, K. *et al.* (2003). Drug/substance reversal effects of a novel tri-substituted benzoflavone moiety (BZF) isolated from *Passiflora incarnata* Linn. – a brief perspective. *Addi Biol* 8 (4): 379-86.

Dhawan, K. *et al.,* (2004). Passiflora: a review update. *J Ethnopharmacol* 94 (1): 1-23.

Dhawan, K.; Kumar, S. and Sharma, A. (2001). Anxiolytic activity of aerial and underground parts of *Passiflora incarnata. Fitoterapia* 72 (8): 922-6.

Dhawan, K.; Kumar, S. and Sharma, A. (2002). Beneficial effects of chrysin and benzoflavone on virility in 2-year-old male rats. *J Food Med* 5 (1): 43-8.

Dhawan, K.; Kumar, S. and Sharma, A. (2003). Antiasthmatic activity of the methanol extract of leaves of *Passiflora incarnata. Phytother Res* 17 (7): 821-2.

Dhawan, K.; Kumar, S. and Sharma, A. (2003). Aphrodisiac activity of methanol extract of leaves of *Passiflora incarnata* Linn in mice. *Phytother Res* 17 (4): 401-3.

Fisher, A.A. *et al.,* (2000). Toxicity of *Passiflora incarnata* L. *J Toxicol Clin Toxicol* 38 (1): 63-6.

Kashani, N. *et al.,* (2001). Passionflower in the treatment of opiates withdrawal: a double-blind randomized controlled trial. *J Clin Pharm Ther* 26 (5): 369-73.

Menghini, A. *et al.,* (1993). Flavonoid Contents in *Passiflora* spp. *Pharmacol Research* 27 (1 1): 13-14.

Neu, R. (1954). Composition of *Passiflora incarnata*. III. The alkaloid of Passiflora, 3-methyl-4-carbolin (2'-methyl-(pyridino-3', 4': 2, 3-indole)). *Arzneimittelforschung* 6 (2): 94-98.

Neu, R. (1954). The constituents of *Passiflora incarnata*. II. Basic components. *Arzneimittelforschung* 4 (10): 601-606.

Oga, S. *et al.,* (1984). Pharmacological Trials of Crude Extract of *Passiflora alata. Planta Medica* 51 (4) 303-306.

Soulamin, R *et al.,* (1997). Behavioural effects of *Passiflora incarnata* L. and its indole alkaloid and flavonoid derivatives and maltol in the mouse. *J Ethnopharmacol* 57 (1): 11-20.

Speroni, E. and Minghetti, A. (1988). Neuropharmacological activity of extracts from *Passiflora incarnata. Planta Med* 488-491.

PATHA (VELVET-LEAF)

Botanical Name: *Cissampelos pareira* Linn.

Natural Order: Menispermaceae

Active Ingredeints: Alkaloids: pelosine or cissampariene, berberine, hyatine, hyatinine, and pareirubrines A and B, grandirubrine and isomerubrine.

Pre-Clinical Studies: Anticancer.

BIBLIOGRAPHY

Mouta et al. (1993). Structures and solid state tantomoac forms of two novel antilubemic alkaloids. *Chem Pharm Bull* (Tokyo) 41(8): 1418-1422.

PUNARNAVA

Botanical Name: *Boerhaavia diffusa* (L) Hk.f. Syn *Boerhaavia repens* Linn.

Natural Order: Nyctaginaceae

English Name: Spreading hog weed.

Active Ingredeints: Alkaloid; punarnavine (Agarwal and Dutt 1935; Nandi and Chatterjee 1974), glycoside: punarnavoside (Seth *et al* 1986; Jain and Khana 1989), rotenoids; boeraviones A-E (Kadota *et al* 1988, Kadota *et al* 1989, Lami *et al* 1990), dihydroisofuraxanthone; boravine (Ahmad and Chung-Ping 1992), lignan: liridodendrin and flavonoids (Pandey *et al* 2005).

Pre-Clinical Studies: Adaptogen (Mugantiwar *et al* 1997). analgesic (Hiruma-Lima *et al* 2000), antibacterial, anticancer (Leyon *et al* 2005), anticonvulsant (Adesina 1979), antidiabetic (Pari, Amarnath Satheesh 2004), antifibrinolytic (Jain and Khana 1989), antihypertensive or calcium-channel blocker (Lami *et al* 1991), anti-inflammatory (Bhalla *et al* 1968; Mudgal 1974), antioxidant (Sateesh, and Pari 2004), antiviral (Verma and Awasthi 1979), cardio-tonic, diuretic (Chowdhury and Sen 1955; Bhide *et al* 1958; Mudgal 1974; Singh 1991), hepatoprotective (Chakraborti and Handa 1989; Chandan, Sharma and Anand 1991; Rawat, *et al* 1994). immunomodulator (Mehrotra *et al* 2002) and immunosuppressant (Pandey *et al* 2005).

Human Studies

In a double blind clinical study done in forty-three healthy adults, tablets of *Boerhavia diffusa* (containing powder of the entire plant) resulted in moderate decrease in thy mean level of blood urea and serum acid phosphatase (no author listed).

Adverse Drug Reactions: Vomiting (emetic nature of the plant).

Teratogenicity: Devoid of teratogenic effect (Singh *et al* 1991).

BIBLIOGRAPHY

Agarwal, R.R. and Dutt, S (1935). Chemical investigation of Punarnva or *Boerhaavia diffusa* Linn. Part II, The isolation of an alkaloid 'Punarnavine' *Proc U Pacad Sci* 5: 240-242.

Ahmad, B and Chung-Ping, Yu (1992). Boravine, a dihydroisofuranoxanthone from *Boerhaavia diffusa. Phytochemistry*, 31 (12) 4382-4384.

Bhalla, T.N. *et al.* (1968). Anti-inflammatory activity of *Boerhaavia diffusa. Indian J Physiol Pharmacol* 12 : 37.

Bhide, N.K. *et al.* (1981). Potassium diuretics in the Ayurvedic system of medicine. *J Postgrad Med* 4 (1): 21-27.

Chakraborti, K.K. and Handa, S.S (1989). Antihepatotoxic activity of *Boerhaavia diffusa. Indian Drugs* 27 (3): 161-166.

Chandan, B.K., Sharma, A.K. and Anand, KK (1991). *Boerhaavia diffusa*: a study of its hepatoprotective activity. *J Ethnopharmacol* 31 (3): 299-307.

Datta, B.K.; Ahmed, M. and Rouf, A.S.S (1992). Rotenoids from *Boerhaavia repens*. International Seminar-Traditional medicine. Calcutta. Abstr. PP. 147.

Hiruma-Lima, C.A. *et al.* (2000). The juice of fresh leaves of *Boerhaavia diffusa* L. (Nyctaginaceae) markedly reduces pain in mice. *J Ethnopharmacol* 71 (1-2): 267-74.

Jain, G.K. and Khanna, N.M. (1989). Punarnavoside, a new antifibrinolytic agent from *Boerhaavia diffusa* Linn. *Indian Journal of Chemsitry* 28 B(2): 163-166.

Kadota, S. *et al.* (1988). Boeravinone A and B, new rotenoid analogue from *Boerhaavia diffusa* Linn. *Chem Pharm Bull* 36: 834-836.

Lami, N. *et al.* (1991). Constituents of the roots of *Boerhaavia diffusa* L. III. Identification of Ca2+ channel antagonistic compound from the methanol extract. *Chem Pharm Bull* (Tokyo) 39 (6): 1551-5.

Lami, N. *et al.* (1991). Constituents of the roots of *Boerhaavia diffusa* L. *IV*-Isolation and Structure determination of boeravinones D, E and F, Chemical and *Pharmaceutical Bulletin* 39 (7): 317-334.

Leyon, P.V., Lini, C.C. and Kuttan, G. (2005) Inhibitory effect of *Boerhaavia diffusa* on experimental metastasis by B16F10 melanoma in C57BL/6 mice. *Life Sci* 76 (12): 1339-49.

Mehrotra, S. *et al.* (2002). Immunomodulation by ethanolic extract of *Boerhaavia diffusa* roots. *Int Immunopharmacol* 2 (7): 987-96.

Mudgal, V. (1974). Comparative studies on the anti-inflammatory and diuretic action with different parts of the plant *Boerhaavia diffusa* Linn. (Punarnava). *J Res Indian Med* 9 (2): 57-59.

Muganntiwar, A.A. *et al.* (1997). Adaptogenic activity of aqueous extract of the roots of *Boerhaavia diffusa* Linn. *Indian Drugs* 34: 184-189.

Nandi, R.P. and Chatterjee, S.K. (1974). Occurrence of punarnavines in *Boerhaavia repens*. *Indian J Exp Biol* 12 (6): 509-511.

Pandey, R. *et al.* (2005). Immunosuppressive properties of flavonoids isolated from *Boerhaavia diffusa* Linn. *Int Immunopharmacol* 5(3): 541-53.

Pari, L. and Amarnath Satheesh, M. (2004). Antidiabetic effect of *Boerhavia diffusa*: effect on serum and tissue lipids in experimental diabetes. *J Med Food* 7 (4): 472-6.

Rawat, A.K. *et al.* (1997). Hepatoprotective activity of *Boerhaavia diffusa* L. roots- a popular ethno medicine. *J Ethnopharmacol* 56 (1): 61-6.

Satheesh, M.A. and Pari, L. (2004) Antioxidant effect of *Boerhavia diffusa* L. in tissues of alloxan induced diabetic rats. *Indian J Exp Biol.* 42 (10): 989-92.

Singh, A. *et al.* (1991). An experimental evaluation of possible teratogenic potential in *Boerhaavia diffusa* in Albino rats. *Planta Med* 57 (4): 315-6.

PURPLE CONE FLOWER

Botanical Name: *Echniacea purpurea* Linn.

Natural Order: Asteraceae

Active Constituents: Alkamides, polysaccharides, peptides, pyrrolizidine alkaloids, caffeic acid derivatives: echinacoside, cynarin, cholrgenic acid.

Non-Clinical Studies: Immunomodulator (Stimpel *et al* 1984; Melchart *et al* 1995).

Clinical Studies: Clinical studies have demonstrated promising effect of purple cone flower in the treatment of upper respiratory infections and common cold (Melchart *et al* 1998; Barrett, Vohmann and Calabrese 1999; Frank 2000; Lindenmuth and Lindenmuth 2000; A trial reported ineffectiveness of purple cone flower for prevention of experimental rhinovirus cold (Turner, Riker and Gangemi 2002).

Adverse Drug Reactions: Recurrent erythema nodosum (Soon and Crawford 2001).

Pregnancy: A study conducted on two hundred six pregnant females demonstrated that use of purple coneflower during pregnancy is not associated with an increased risk for major malformations (Gallo *et al* 2000).

Toxicity: LD 50 for fresh juice in rats and mice: 5 and 10 g/kg body weight i.v., 15 and 30 g/kg body weight orally (European Agency for Evaluation of Medicinal Products).

BIBLIOGRAPHY

Barrett. B.; Vohmann, M. and Calabrese, C. (1999). Echinacea for upper respiratory infection. *J Fam Pract* 48: 628-635.

Frank, L.G. (2000). The efficacy of Echinacea compound herbal tea preparation on the severity and duration of upper respiratory and flu symptoms: a randomized, double blind, placebo-controlled study. *J Comp Alt Med* 6 (4): 327-334

Gallo, M. *et al.,* (2000). Pregnancy outcome following gestational exposure to echinacea. *Arch Intern Med* 160: 3141-3143.

Lindenmuth, G.F. and Lindenmuth, E.B. (2000). The efficacy of echinacea compound herbal tea preparation on the severity and duration of upper respiratory and flu symptoms: a randomized, double-blind placebo-controlled study. *J Altern Complement Med* 6 (4): 327-334

Melchart, D. *et al.* (1998). Echinacea root extracts for the prevention of upper respiratory tract infections: a double-blind, placebo-controlled randomized trial. *Arch Fam Med* 7: 541–545.

Melchart, D, *et al.* (1995). Results of five randomized studies on the immunomodulatory activity of preparations of echinacea. *J Alt Comp Med* 1 (2): 145–160.

Mullins, R.J. and Heddle, R. (2002). Adverse reactions associated with echinacea: the Australian experience. *Ann Allergy Asthma Immunol* 88 (1): 42-51

Soon and Crawford (2001). Recurrent erythema nodosum associated with Echinacea herbal therapy. *J Am Acad Dermatol* 44 (2): 298-299.

Stimpel, M. *et al.* (1984). Macrophage activation and induction of macrophage cytotoxicity by purified polysaccharide fractions from the plant *Echinacea purpurea. Infect Immunity* 46: 845-849.

Turner, R.B.; Riker, D.K. and Gangemi, J.D. (2000). Ineffectiveness of Echinacea for prevention of experimental rhinovirus colds. *Antimicrob Agents Chemother* 44: 1708-1709.

PUSHKARMOOL (ELECAMPANE)

Botanical Name: *Inula racemosa* Hook f.

Natural Order: Asteraceae.

Active Ingredients: Inulin, aromatic oil containing alantolactone or helenalin and alkaloid: saussurine.

Non-Clinical Studies: Anti-anginal (Tripathi, Upadhyaya and Gupta 1984), adrenergic beta-blocking (Tripathi *et al* 1988) and hypoglycemic (Tripathi and Chaturvedi 1995).

BIBLIOGRAPHY

Tripathi, Y.B. and Chaturvedi, P. (1995), Assessment of endocrine response to *Inula racemosa* in relation to glucose homeostasis in rats. *Indian J Exp Biol* 33: 686-9.

Tripathi, Y.B. *et al.,* (1988). Assessment of the adrenergic beta-blocking activity of *Inula racemosa. J Ethnopharmacol* 23 (1): 3-9.

Tripathi, Y.B., Upadhyaya. and Gupta, V.K. (1984). Beneficial effect of *Inula racemosa* (pushkarmoola) in angina pectoris: *Indian Journal of Physiol Pharmacol* A preliminary report 28 (1): 73-5.

RAJARKA

Botanical Name: *Calotropis gigantea* R. Br

Natural Order: Asclepidaceae.

Active Ingredients: Cardiac glycosides: calotropine, principle of peculiar nature: mudarine.

Non-Clinical Studies: Anti-diarrhoeal (Chitme *et al* 2004) and anthelmintic.

BIBLIOGRAPHY

Chitme, H.R. *et al.* (2004). Studies on anti-diarrhoeal activity of *Calotropis gigantea* R.Br. in experimental animals. *J Pharm Pharm Sci* 7(1): 70-5.

ROMAN CHAMOMILE

Botanical Name: *Chamaemelum nobile* (Linn.) Allioni Syn. *Anthemis nobilis* Linn.

Natural Order: Asteraceae

Active Constituents: Sesquiterpenes: nobilin, 3-epinibilin, 1, 10-epoxynobili, 3-dehydronobilin (Anon: PDR for Herbal Medicines). flavonoids, polyacetylenes and triterpenes.

Non-Clinical Studies: Antibacterial (Anon: PDR for Herbal Medicines) and cytostatic (Issac 1993).

BIBLIOGRAPHY

Anon: PDR for Herbal Medicines. Medical Economics Company Inc, Montvale, NJ; 2000: 812-813.

Issac, O, (1993). *Chamaemelum nobile* (L.) Allioni. *ZPT* 14 (4): 212.

Foster, S. and Duke, J.A. (1990). Chamomile in Medicinal Plants. Houghton Mifflin Co., New York, NY, p. 84.

ROUNDLEAF SUNDEW

Botanical Name: Drosera rotundifolia L.

Natural Order: Droseraceae.

Active Ingredients: Napthoquinone: plumbagin (Budzianowski *et al* 1993), 7-methyljuglone (Kamarienen *et al* 2003), hydro-plumbagin glucoside (Vikenborg *et al* 1969); glycoside: rossoliside (Vikenborg *et al* 1970) and flavonoids (Bienenfeld and Katzlmeier 1966).

Non-Clinical Studies: Anti-tumor (Budzianowski *et al* 1993).

BIBLIOGRAPHY

Budzianowski, J. *et al.* (1993). Ellagic acid derivatives and futher naptho-quinones from *Dinoea muscipula* and four species of Drosera *In vitro* cultures. *Planta Med* 59 (7): A 654.

Vikenborg, J. *et al.* (1969). The presence of hydroplumbagin glucoside in *Drosera rotundifolia* L. *Pharm Wekbl* 104 (3): 45-9.

Vikenborg, J. *et al.* (1970). Rossoliside, a glycoside from *Drosera rotundifolia* L. *Pharm Wekbl* 105 (14): 414.

Bienenfeld W. and Katzlmeier, H (1966). Flavonoids from *Drosera rotundifolia* L. *Arch Pharm Ber Dtsch Pharm Ges* 229 (7): 598-602.

Kamarienen, T. *et al.* (2003). Regional and habitat differences in 7-methyljuglone content of Finnish *Drosera rotundifolia Phytochemistry* 63 (3): 309-14.

SALAI-GUGGUL

Botanical Name: *Boswellia serrata* Roxb. ex Colebr. Syn *Boswellia glabra* (Roxb.) Bennet, *Boswellia glabra* Roxb.

Natural Order: Burseraceae

English Name: Indian Olibanum tree.

Active Ingredeints: Tetracyclic triterpene acids including α-boswellic acid and its 11-keto derivatives and their acetates, ß-boswellic acid, 3-O-acetyl-beta-boswellic acid, 11-keto-beta-boswellic acid and 3-O-acetyl-11-keto-beta-boswellic acid (Beton *et al* 1956; Krohn *et al* 2001). Pardhy and Bhattacharya 1978), diterpene alcohol; serratol (Pardh and Bhattacharya 1978) and ß-sitosterol (Beri and Karnik, 1963).

Pre-Clinical Studies: Analgesic (Karbn and Menon 1969; Menon 1969), anti-atherosclerotic (Zutshi *et al* 1986). antibact-erial, anticancer (Huang *et al* 2000; Shoy *et al* 1998; Xia *et al* 2005; 4:381-388 {boswellic acetate}), antifungal, anti-inflammatory (Ammom *et al* 1993; Safayhi *et al* 1994; Singh and Atal 1986; Wildfeuer Neu and Safayhi, 1998), anti-arthritic, hypotensive (Kar 1977). immunomodulator (Sharma *et al* 1996) and sedative.

Human Studies

Rheumatoid arthritis

1. Two placebo controlled studies, involving a total of 81 people with rheumatoid arthritis, reportedly found significant reductions in swelling and pain over the course of 3 months with *Boswellia serrata* (Etzel, 1996).
2. A recent double-blind placebo-controlled study that enrolled 78 patients found no benefit. About half of the patients dropped out, which diminishes the signifi-cance of the results (Sander, Herborn and Rau 1998).

Bronchial asthma

A 6-week double-blind placebo-controlled study of 80 individuals with relatively mild asthma found that treatment with *Boswellia serrata* mg 3 times daily reduced the frequency of asthma attacks and improved objective measurements of breathing capacity (Gupta *et al* 1998)

Ulcerative colitis

Animal studies have shown beneficial effect of acetyl-11-keto-beta-boswellic on experimental ileitis. (Krieqlstein *et al* 2001).

A small clinical study found that people with ulcerative colitis taking 550 mg of *Boswellia serrata* gum resin three times daily for six weeks had similar improvement in symptoms and the severity of their disease as people with ulcerative colitis taking the drug sulphasalazine. Overall, 82% of patients receiving *Boswellia serrata*, along with 75% of z-blind, verum-controlled, parallel group study in 102 patients demonstrated efficacy of *Boswellia serrata* extract in the treatment of regional ileitis. 44 patients were treated with *Boswellia serrata* extract and 39 patients treated with mesalazine. The study confirms that therapy with *Boswellia serrata* extract was not inferior to mesalazine. Further taking into consideration safety and efficacy *of Boswellia serrata* extract it appears to be superior over mesalazine in terms of a benefit-risk-evaluation (Gerherdat *et al* 2001).

Toxicity: No literature.

BIBLIOGRAPHY

Ammom, H.P.T. *et al.* (1993). Mechanism of anti-inflammatory actions of curcumin and boswellic acid. *J Ethnopharmacology* 38 (2-3): 113-119.

Beri, R.M. and Karnik, M.G. (1963). Phytosterol: ß-sitosterol from *Boswellia serrata* Roxb. *Curr Sci* 32: 324.

Beton, J.L. *et al.* (1956). The chemistry of triterpenes and related compound. Part-XXV111 ß boswellic acid. *J Chem Soc* 2904-2909.

Etzel, R. (1996). Special extract of Boswellia serrata (H 15) in the treatment of rheumatoid arthritis. *Phytomedicine* 3: 91-94.

Gerherdat, H., *et al.* (2001). Therapies of active Crohn disease with *Boswellia serrata* extract H 15. *Z Gastroeneterol* 39 (1): 11-7.

Gupta, I., *et al.* (1998). Effects of *Boswellia serrata* gum resin in patients with bronchial asthma: results of a double-blind, placebo-controlled, 6-week clinical study. *Eur J Med Res* 3: 511-514.

Gupta, I., *et al.* (1997). Effect of *Boswellia serrata* gum resin in patients with ulcerative colitis. *Eur J Med Res* 2: 37-43.

Huang, M.T., *et al.,* (2000). Anti-tumor and anti-carcinogenic activities of triterpenoid, beta-boswellic acid. *Biofactors* 13 (1-4): 225-30.

Kar, A. and Menon, M.K. (1969). Analgesic effect of the gum-resin of *Boswellia serrata, Life Science* 8 (1): 1023-1028.

Kar, A. (1977). Effect of the gum-resin of *Boswellia serrata* Roxb. on the cardiovascular system isolated tissues, *Ind Drug Pharm Ind* 12 (4): 17-20.

Krohen K, *et al.*, (2001). *Phytochem Ana* 12 (6): 374-6.

Menon, N.K. (1969) Analgesic effect of the gum resin of *Boswellia serrata* Roxb. *Life Sci* 8: 1023.

Pardhy, R.S. and Bhattacharya, S.C. (1978). Structure of serratol, a new diterpene cembranoid alcohol from *Boswellia serrata* Roxb *Indian J Chem* 16 B: 171-173.

Pardhy, R.S. and Bhattacharya, S.C. (1978). ß-boswellic acid, acetyl- ß-boswellic acid, acetyl-II-keto- ß-boswellic acid and II. keto= ß-boswellic acid, four pentacyclic triterpene acids from resin of *Boswellia serrata* Roxb., *Indian J Chem* 16B: 176-178.

Safayhi, H *et al.,* (1994). Structure requirements for 5-Lo inhibition by boswellic acids, *European Journal of Pharmaceutical Sciences* 2 (1-2): 101.

Sander, O., Herborn, G., Rau, R., (1998). Is H15 (resin extract of *Boswellia serrata,* "incense") a useful supplement to established drug therapy of chronic polyarthritis? Results of a double-blind pilot study [in German, English abstract]. *Z Rheumatol* 57: 11–16.

Sharma, M.L. *et al.,* (1996). Immunomodulatory activity of boswellic acids (pentacyclic triterpene acids) from *Boswellia serrata. Phytotherapy Research* 10 (2) 107-112.

Shoy *et al.,* (1998). Inhibitory activity of boswellic acids from *Boswellia serrata* against human leukemia HL-60 cells in culture. *Planta Med* 64 (4): 328-31.

Singh, G.B., Atal, C.K., (1986). Pharmacology of an extract of salai guggal ex-*Boswellia serrata,* a new non-steroidal anti-inflammatory agent. *Agents Actions* 18: 407–412.

Wildfeuer, A., Neu, I.S., Safayhi, H., *et al.,* (1998). Effects of boswellic acids extracted from an herbal medicine on the biosynthesis of leukotrienes and the course of experimental autoimmune encephalomyelitis. *Arzneimittelforschung* 48: 668–674.

Zutshi, U. *et al.,* (1986). Mechanism of cholesterol lowering effect of Salai guggul ex-*Boswellia serrata* Roxb., *Indian J. Pharmacology* 18(3) 182-183.

SAPTACHAKRA

Botanical Name: *Salacia chinensis* Linn.

Natural Order: Hippocrateaceae.

Active Ingredients: Salacinol (Ghavami, Johnston and Pinto 2001; Yoshikawa *et al* 2003) and proanthocyanidins (Krishnan and Rangaswamy 1967).

Non-Clinical Studies: Alpha glucosidae inhibitor (Yoshikawa *et al* 2003) and hypotensive (Jansakul, Jusapalo and Mahattanadul).

BIBLIOGRAPHY

Ghavami, A.; Johnston, B.B. and Pinto, B.M. (2001). A new class of glycosidase inhibitor: synthesis of salacinol and its stereoisomers. *J Org Chem* 66 (7): 2312-7.

Jansakul, C.; Jusapalo, N. and Mahattanadul, S. (1998). Hypotensive effect of N-butanol extract from stem of *Salacia chinensis* in rats. ISHS Acta Horticulturae 678: III WOCMAP Congress on Medicinal and Aromatic Plants- Vol. 4: Targeted Screening of Medicinal and Aromatic Plants, Economics and Law.

Krishnan, V. and Rangaswamy, S. (1967). Proanthocyanidins of *Salacia chinensis* Linn. *Tetrahedron Lett* 26: 2441-6.

Yoshikawa, M. *et al.* (2003). Biological activities of *Salacia chinensis* originating in Thailand: the quality evaluation guided by alpha-glucosidase inhibitory activity. *Yakaugaku Zasshi* 123 (10): 871-80.

SARIVA

Botanical Name: *Hemidesmus indicus* R.Br.

Natural Order: Asclepidaceae.

Englsih Name: Indian Sarsaparilla

Active Ingredients: Coumarin, volatile oil (containing p-methoxy salicylic aldehyde). pregnane oligoglycosides: medidesmine, hemisine, desmisine, denicunine and heminine (Deepak Srivastava, and Khare 1997; Sigler *et al* 2000) and pentacyclic triterpenes (Roy *et al* 2001).

Non-Clinical Studies: Antileprotic (Gupta 1981), hepatoprotective (Prabakan Anandan and Devaki 2000), chemopreventive (Sultana *et al* 2003), antidiarrhoeal (Das, Prakash, and Devaraj (2003). antioxidant and antithrombotic (Mary*et al* 2003). renoprotective (Kotnis *et al* 2004).

BIBLIOGRAPHY

Das, S.; Prakash, R. and Devaraj, S.N. (2003). Antidiarrhoeal effects of methanolic root extract of *Hemidesmus indicus* (Indian sarsaparilla)--an *In vitro* and *in vivo* study. *Indian J Exp Biol* 41 (4): 363-6.

Deepak, D.; Srivastava, S. and Khare, A. (1997). Pregnane glycosides from *Hemidesmus indicus. Phytochemistry* 44 (1): 145-51.

Gupta, P.N. (1981). Antileprotic action of an extract from "Anantamul' (*Hemidesmus indicus* R. Br.). *Lepr India* 53 (3): 354-9.

Kotnis, M.S. *et al.* (2004). Renoprotective effect of *Hemidesmus indicus*, an herbal drug used in gentamicin-induced renal toxicity. *Nephrology* 9 (3):142-52.

Mary, N.K. *et al.* (2003). *In vitro* antioxidant and antithrombotic activity of *Hemidesmus indicus* (L) R.Br. *J Ethnopharmacol* 87 (2-3): 187-91.

Prabakan, M.; Anandan, R. and Devaki, T. (2000). Protective effect of *Hemidesmus indicus* against rifampicin and isoniazid-induced hepatotoxicity in rats. *Fitoterapia* 71 (1): 55-9.

Roy, S.K. *et al.* (2001). New pentacyclic triterpenes from the roots of *Hemidesmus indicus. Pharmazie.* 56(3):244-6.

Sigler, P. *et al.* (2000). A. C21 steroidal glycosides from *Hemidesmus indicus. Phytochemistry* 54 (8): 983-7.

Sultana, S. *et al.* (2003). Inhibition of cutaneous oxidative stress and two-stage skin carcinogenesis by *Hemidesmus indicus* (L.) in Swiss albino mice. *Indian J Exp Biol* 41(12): 1416-23.

SCARLET PIMPERNEL

Botanical Name: *Anagallis arvensis* Linn.

Natural Order: Primulaceae

Active Constituents: Triterpene saponin: anagalline, apoanagallosaponins III and IV (Aliotta *et al* 1992; Shoji *et al* 1997). triterpene glycosides: anagalloside, anagallosaponins VI-IX, cucurbitacins E, B, D, I and L (Miro 1995) and flavonoids.

Non-Clinical Studies: Anthelmintic, antifungal, anti-spermatogenic, antiviral (Amoros, Fauconnier and Girre 1992), cytotoxic, ecbolic, estrogenic and hemolytic, molluscicidal (Abdel-Gawad *et al* 2000).

Adverse Drug Reations: Gastroenteritis and nephritis.

Toxicity: Anorexia, restlessness, diarrhea, thirst, difficult breathing, tremors and ended by coma, death, reduction in PCV%, Hb concentration and RBCs count, increase in urea and creatinine. LD_{50} of alcoholic extract: 10.718 mg/kg.b.wt. (Sultan, Hussein and Hegazy 2003)

BIBLIOGRAPHY

Abdel-Gawad, M.M. *et al.* (2000). Molluscicidal saponins from *Anagallis arvensis* against schistosome intermediate hosts. *Jpn J Infect Dis* 53 (1): 17-9.

Aliotta, G. *et al.* (1992). An oleanen triterpene from *Anagallis arvensis*. *PH* 1 (3): 929-933.

Amoros, M.; Fauconnier, B. and Girre, R.L. (1992). *In vitro* antiviral activity of a saponin from *Anagallis arvensis*, Primulaceae, against herpes simplex virus and poliovirus. *Antiviral Res* 8: 13-25.

Miro, M. (1995). Cucurbitacins and their pharmacological effects, *Phytotherapy Research* 9 (3). 159-168.

Shoji, N. *et al.* (1997). Triterpenoid glycosides from *Anagallis arvensis*. *Phytochemistry* 37 (5): 1397-402.

Sultan, I.A.L.; Hussein, Y.A. and Hegazy, A. (2003). Toxicity of *Anagallis arvensis* Plant. *Pakistan Journal of Nutrition* 2 (3): 116-122,

SCOTH BROOM

Botanical Name: *Cytisus scoparius* (Linn.). Syn *Spartium scoparium* (Linn.).

Natural Order: Fabaceae.

Active Ingredients: Alkaloid: sparteine (Koldoziejski *et al* 1964), isoflavonoids; glucoside: scoparin, lectin (Konami *et al* 1992).

Non-Clinical Studies: Cardiac stimulant.

BIBLIOGRAPHY

Koldoziejski, J. *et al.* (1964). Localization of sparteine in *Cytisus scoparius* Linn. during the vegetation stage. *Acta Pol Pharm* 21: 501-8.

Konami, Y. *et al.* (1992). The primary structure of the *Cytisus scoparius* seed lectin and a carbohydrate-binding peptide. *J Biochem* (Tokyo) 112 (3): 366-75.

SEA BUCKTHORN

Botanical Name: *Hippophae rhamnoides* Linn.

Natural Order: Elaeagnaceae.

Active Ingredients: Vitamin: ascorbic acid, flavonoids, fatty oil containing essential fatty acids, linoleic acid and a-linolenic acid, polyunsaturated fatty acids: arachidonic acid and eicosapentaenoic acids.

Non-Clinical Studies: Anti-ageing, gastroprotective (Suleyman, *et al* 2001).

Clinical Studies

Atopic dermatitis

Forty-nine patients received a daily dose of 5 g *Hippophae rhamnoides* seed oil pulp oil or paraffin oil for four months. The fatty acid compositions of plasma phospholipids and neutral lipids of the patients were analyzed before, after one month and at the end of the treatment. Skin biopsies were taken from sixteen patients, and the fatty acid compositions of skin glycerophospholipids were analyzed before and after the treatment. Supplementation of the seed oil significantly increased the proportion of -linolenic acid in plasmaneutral lipids and total n-3 fatty acids in plasma phospholipids and neutral lipids. Increases in the proportion of a-linolenic acid, linoleic acid and eicosapentaenoic acid in plasma phospholipids by the seed oil treatment were close to significant. The pulp oil treatment increased the proportion of palmitoleic acid and lowered the proportion of pentadecanoic acid in both plasma phospholipids and neutral lipids. Seed oil treatment slightly increased the level of docosapentaenoic acid and decreased the level of palmitic acid in skin glycerophospholipids. These results indicate a higher efficiency of incorporation and metabolism of a-linolenic acid than than linoleic acid and a relatively stable fatty acid composition of skin glycerophospholipids (Baoru, Kallio and Tahvonen 2002).

BIBLIOGRAPHY

Baoru, Y.; Kallio, H. and Tahvonen, R. (2002) Effects of Dietary Supplementation of Seabuckthorn (*Hippophae rhamnoides*) Oils on Fatty Acids in Patients with Atopic Dermatitis. *J Nitr Biochem* 13(6): 346-354.

Suleyman, H. *et al.* (2001). The effects of *Hippophae rhamnoides* L. extract on ethanol-induced gastric lesion and gastric tissue glutathione level in rats: A comparative study with melatonin and omeprazole. *Indian Journal of Pharmacology* 33 (2): 7-81.

SHALPARNI

Botanical Name: *Desmodium gangeticum* DC.

Natural Order: Fabaceae.

Active Ingredients: Alkaloid, ash, resin, pterocarpenoid: gangetin (Ghosh and Anadakumar 1983).

Non-Clinical Studies: Anti-inflammatory, analgesic (Ghosh and Anadakumar 1983).

BIBLIOGRAPHY

Ghosh, D, and Anadakumar, A. (1983). Anti-inflammatory and analgesic activities of Gangctin-a pterocarpenoid from *Desmodium gangeticum*. *Ind J Pharmacol* 15(4): 391-462.

SHARPAUNKHA (WILD INDIGO)

Botanical Name: *Tephrosia purpurea* (L.) Pers.

Natural Order: Fabaceae.

Active Ingredients: Colouring matter: tephrosin, bioflavonoid: rutin, rotenoid.

Non-Clinical Studies: Hepatoprotective (Ramamurthy and Srinivisan 1993).

BIBLIOGRAPHY

Ramamurthy, M.R. and Srinivisan, M. (1993). Hepatoprotective effect of *Tephrosia purpurea* in experimental animals. *Indian Journal of Pharmacology* 25:34-36.

SHATAVARI

Botanical Name: *Asparagus racemosus* Willd.

Natural Order: Liliaceae

Active Ingredients: Polycyclic alkaloid: aspargamine-A (Sekine et *al* 1994), disogenin (Nair and Subramanian 1969), sarsapogenin (Kar and Sen 1985; Oketch-Rabah 1998), saponins: shatavarin 1 to 4 (Joshi 1988; Shao *et al* 1997; Joshi and Dev 1988), isoflavones; 8-methoxy-5, 6, 4'-trihydroxyiso-flavone 7-O-beta-D-glucopyranoside (Saxena and Chourasia 2001), 9, 10-dihydrophenanthrene: racemosol (Sekine 1997), disaccharide (Landge and Bose 1970).

Non-Clinical Studies: Antibacterial (Dhankur *et al* 1986; Mandal *et al* 2000), antihepatotoxic (Muruganadan *et al* 2000), anti-oxytocic (Gaitonde and Jetmalan 1969), anti-ADH (Gaitonde and Jetmalani 1969), prokinetic (Dalvi *et al* 1990), anticancer (Rao 1981; Sekine. 1997; Shao *et al* 1997), antitussive (Mandal *et al* 2000). antioxidant (Kamat *et al* 2000), digestive (Dange Kanitkar and Pendse 1969), galactogouge (Joglekar *et al* 1967), immunomodulator (Rege and Dahanukar 1993), gastroprotective (Dhankur *et al* 1983).

Clinical Studies

Failing lactation

A study has been done to evaluate efficiacy of *Asparagus officinalis* in failing lactation but details are missing (Sharma *et al* 1996).

Duodenal ulcer

A study has been done to evaluate efficiacy of *Asparagus officinalis* in duodenal ulcer but details are missing (Kishore *et al* 1980).

BIBLIOGRAPHY

Dahanukar S.A. *et al.,* (1983). Cytoprotective effect of *Terminalia chebula* and *Asparagus racemosus* on gastric mucosa. *Indian Drugs* 21: 442-5.

Dahanukar S.A *et al.,* (1986). Protective effect of *Asparagus racemosus* against induced abdominal sepsis. *Indian Drugs* 24 (3): 125-128.

Dalvi, S.S. *et al.,* (1990). Effect of *Asparagus racemosus* (*Shatavari*) on gastric emptying time in normal healthy volunteers. *J. Postgrad Med* 36 (2): 91-4.

Dange, P.S.; Kanitkar, U.K. and Pendse, G.S. (1969). Amylase and Lipase activites in the root of *Asparagus racemosus. Planta Medica* 17(4): 393-395.

Gaitonde, B.B. and Jetmalani, M. (1969). Anti-oxytocic action of saponin isolated from *Asparagus racemosus* Willd. (Shatavari) on uterine muscle, *Arch Int Pharmacodyn Ther* 179 (1): 121-129.

Gaitonde, B. B. and Jetmalani, M. (1969). Anti-oxytocic and anti-ADH actiivty of saponin isolated from *Asparagus racemosus* Willd. *Indian J Pharm* 31: 175.

Goyal, R.K.; Singh, J. and Lal, H. (2003). *Asparagus racemosus*--an update. *Indian J Med Sci* 57: 408-14.

Joglekar, G.V. *et al.,* (1967). Galactogouge effect of *Asparagus racemosus*, Preliminiary commnucation. *Indian Med J* 61 (7): 165.

Joshi, J.D.S. (1988). Chemistry of ayurvedic crude drugs: Part VIII: Shatavari: 2. Structure elucidation of bioactive shatavarin I and other glycosides. *Indian Journal of Chemistry Section B Organic Chemistry Including Medicinal Chemistry* 27 (1): 12-16.

Joshi, J. and Dev, S. (1988). Chemistry of Ayurvedic crude drugs: Part VIIIa-Shatavari-2: Structure elucidation of bioactive Shatavarin-I and other glycosides b, c. *Indian J Chem* 27B: 12-6.

Kamat, J.P. *et al.,* (2000). Antioxidant properties of *Asparagus racemosus* against damage induced by gamma-radiation in rat liver mitochondria. *Journal Ethnopharmacol* 71: 425-35.

Kar, D.K. and Sen, S. (1985). Sarsasapogenin in callus culture, *Current Science* 549 (12): 585.

Kishore, P. *et al.,* (1980). Pandey PN, Pandey SN, Dash S. Treatment of duodenal ulcer with Asparagus racemosus Linn. *J Res Indian Med Yog Homeo* 15: 409-15.

Landge, A.B. and Bose, J.L. (1970). Structure of Diasaccharide from *Asparagus racemosus, Indian Journal of Chemistry* 8: 588-589.

Mandal S.C. *et al.* (2000). Evaluation of antibacterial activity of *Asparagus racemosus* Willd. root. *Phytother Research* 14: 118-9.

Mandal S.C. *et al.* (2000). Antitussive effect of *Asparagus racemosus* root against sulfur dioxide-induced cough in mice. *Fitoterapia* 71: 686-9.

Muruganadan, S. *et al.* (2000). Studies on the immunostimulant and antihepatotoxic activities of *Asparagus racemosus* root extract. *J Med Arom PI Sci* 22: 49-52.

Nair, A.G.R. and Subramanian, S. (1969). Occurrence of disogenin in *Asparagus racemosus, Current Science* 17: 414.

Oketch-Rabah, H.A. (1998). Phytochemical Constituents of the Genus Asparagus and their biological activities. *Hamdard* 41: 33-43.

Rao, A.R. (1981). Inhibitory action of *Asparagus racemosus* on DMBA-induced mammary carcinogensis in rats. *Int J Cancer* 28 (5): 607-610.

Rege, N.N. and Dahanukar, S.A. (1993). Quantitation of microbicidal activity of mononuclear phagocytes: an *In vitro* technique. *Journal Postgrad Medicine* 39: 22-5.

Saxena, V.K. and Chourasia, S. (2001). A new isoflavone from the roots of *Asparagus racemosus. Fitoterapia* 72: 307-9.

Sekine,T. *et al.* (1994). Structure of Asparagamine-A, a novel polycyclic alkaloid from *Asparagus racemosus. Chem Pharm Bull* 496: 1360-1362.

Sekine. (1997). Racemosol, a 9,10- Dihydrophenanthrene from *Asparagus racemosus. Phytochemistry* 44: 763-4.

Shao, Y.U. *et al.,* (1997). Steroidal saponins from *Asparagus officinalis* and their cytotoxic activity. *Planta Med* 63: 258-62.

Sharma, S. *et al.* (1996). Randomized controlled trial of *Asparagus racemosus* (Shatavari) as a lactogogue in lactational inadequacy. *Indian Pediatr* 33: 675-677.

SHATI

Botanical Name: *Hedychium spicatum* Linn. Buch.-Ham

Natural Order: Zingiberaceae

Active Ingredients: Oil, glucoside, starch, ash, hedychenone.

Non-Clinical Studies: Anti-inflammatory and analgesic.

SHEPHERD'S PURSE

Botanical Name: *Capsella bursa pastoris* (L.) Medik.

Natural Order: Cruciferae

Active Constituents: Cardiac glycosides, volatile oil, polypeptides, flavonoids, alkaloid: bursine, organic acid: bursic acid.

Non-Clinical Studies: Antihypertensive (Anon: PDR for Herbal Medicines), antifertility, ecbolic (Anon: PDR for Herbal Medicines), haemostatic (Vermathen *et al* 1993).

BIBLIOGRAPHY

Anon: PDR for Herbal Medicines. Medical Economics Company Inc, Montvale, NJ; 2000: 812-813.

Vermathen, M. *et al.* (1993). Effect of the herb extract of *Capsella bursa pastoris* on blood coagulation. *Planta Med* 59 (97): A670

SHIGRU (DRUMSTICK PLANT)

Botanical Name: *Moringa oleifera* Lam.

Natural Order: Moringaceae.

Active Ingredients: Essential oil (DanMalam *et al* 2001). alkaloids: moringine (identical to benzylamine), moringinine (The Ayurvedic Pharmacopoeia of India), spirochine, ptergyospermin (Kurup *et al* 1954; Das *et al* 1957), anthomine, resin, mucilage, proanthocyanidin (De Silva Fernando Sirimanne 1986), nitrile glycosides: niazirin and niazirinin, mustard oil glycosides: 4-[(4'-O-acetyl-alpha-L-rhamnosyloxy) benzyl] isothiocyanate, niaziminin A, niaziminin B (Faizi *et al* 1994).

Non-Clinical Studies: Hepatoprotective (Pari and Kumar 2002). hypoglycemic (Bour *et al* 2005), bronchodilator, antifungal (Kurup *et al* 1954), antibiotic (Kurup *et al* 1954; Das *et al* 1957; Das *et al* 1957; Càceres 1991; Càceres *et al* 1991), depressant action on smooth muscle fibers, cardiac stimulant, antibiotic, uterine tonic, inhibitory effect on the tone and movements of the intestine, antispasmodic (Càceres *et al* 1992), anti-inflammatory (Càceres *et al* 1992), diuretic (Càceres *et al* 1992), antipyretic, antifertility (Shukla, Mathur and Prakash 1998), estrogenic, anti-estrogenic, progestational, antiprogestational and antitumor (Guevara *et al* 1999).

Toxicity: The LD50 of ptergyospermin in mice and rats, S.C: greater than 350 to 400 mg/kg body weight (Das *et al 1957)*. Moringinine is toxic and acts on the parasympathetic nervous system. Spirochin paralyses nerves.

BIBLIOGRAPHY

Bour, S. *et al.* (2005). Effects of oral administration of benzylamine on glucose tolerance and lipid metabolism in rats. *J Physiol Biochem* 61 (2): 371-380.

Càceres, A. *et al.* (1991). Pharmacological properties of *Moringa oleifera*. 1: Preliminary screening for antimicrobial activity. *Journal of Ethnopharmacology* 33: 213-216.

Càceres, A. *et al.* (1992). Pharmacologic properties of *Moringa oleifera*. 2: Screening for antispasmodic, anti-inflammatory and diuretic activity, *Journal of Ethnopharmacology* 36: 233- 237.

Càceres, A. (1991). Pharmacological properties of *Moringa oleifera*. 3. Effect of seed extracts in the treatment of experimental pydermia, *Fitoterapia* LXII (5): 449-450.

DanMalam, H.U. *et al.* (2001). Pharmacognisitic studies on leaves of *Moringa oleifera. Nigerian Journal of Natural Products* 5

Das, B.R. *et al.* (1957). Antibiotic principle from *Moringa pterygosperma*. VIII. Some pharmacological properties and *in vivo* action of pterygospermin and related compounds. *Indian J Med Res* 45 (2): 197-206.

Das, B.R. *et al.* (1957). Antibiotic principle from *Moringa pterygosperma*. VII. Antibacterial activity and chemical structure of compounds related to pterygospermin. *Indian J Med Res* 45 (2): 191-6.

De Silva, K.T.D., Fernando, A.C.M. and Sirimanne, V.D.P. (1986). Studies on the chemistry of Moringa Gum (*Moringa oleifera*). Proceedings of the Sri Lanka Association for the Advancement of Science 42 (1): 171.

Faizi, S. *et al.* (1994). Isolation and structure elucidation of new nitrile and mustard oil glycosides from *Moringa oleifera* and their effect on blood pressure. *J Nat Prod* 57 (9): 1256-1261

Guevara, A.P. *et al.* (1999). An antitumor promoter from *Moringa oleifera* Lam. *Mutat Res* 6: 440 (2): 181-188.

Kurup, P.A. *et al.* (1954). Antibiotic principle from *Moringa pterygosperma*. VI. Mechanism of antibacterial action of pterygospermin inhibition of transaminase by pterygospermin. *Indian Journal of Medical Research* 42 (1): 115-23.

Kurup, P.A. *et al.* (1954). Antibiotic principles from *Moringa pterygosperma*. III. Action of pterygospermin on germination of seeds and filamentous fungi. *Indian Journal of Medical Research* 42 (1): 97-9.

Pal, Saroj K., Mukherjee, Pulok K. and Saha, B.P. (1995). Studies on the Antiulcer Activity of *Moringa oleifera* Leaf Extract on Gastric Ulcer Models in Rats, *Phytotherapy Research* 9: 463-465.

Pari, L. and Kumar, N.A. (2002). Hepatoprotective activity of *Moringa oleifera* on antitubercular drug-induced liver damage in rats. *J Med Food* 5 (3):171-7.

Shukla, S.; Mathur, R. and Prakash, A.O. (1998). Antifertility profile of the aqueous extract of *Moringa oleifera* roots. *J Ethno-pharmacol* 22 (1): 51-62

The Ayurvedic Pharmacopoeia of India, Part I, (II) 155

Verma, S.C. *et al.* (1976). Nutritional value of moringa. *Current Sci* 45 (21): 769–770.

SIBERIAN GINSENG

Botanical Name: *Eleutherococcus senticosus* Maxim

Natural Order: Droseraceae.

Active Ingredients: Polysaccharides: eleutherane A-G, triterpene glycosides: eleutheroside A-M, coumarin: isofraxidin, lignans

Non-Clinical Studies: Immunomodulator (Bohn, Nebe and Birr, 1987) and sedative (Medon, Ferguson and Watson 1984).

Clinical Studies

Fatigue

Sixteen subjects were paired and analyzed for 6 weeks on therapy and 2 weeks after discontinuation. No significant difference was documented in any of the indices evaluated including heart rate, oxygenation, serum lactate, time to exhaustion, and psychological measurements. No adverse reactions were noted (Dowling *et al* 1996).

Drug Interactions: Digoxin (McRae 1996).

BIBLIOGRAPHY

Bohn, B., Nebe, C.T. and Birr, C. (1987). Flow Cytometric Studies with *Eleutherococcus senticosus* Extract as an Immuno modulating Agent. *Drug Res* 37 (10): 1193-1196.

Dowling, E.A., *et al.* (1996). Effect of *Eleutherococcus senticosus* on submaximal and maximal exercise performance. *Med Sci Sports Exerc* 28: 482-489.

McRae, S. (1996). Elevated serum digoxin levels in a patient taking digoxin and Siberian ginseng. *CMAJ* 155: 293-295.

Medon, P.J.; Ferguson, P.W. and Watson, C.F. (1984). Effects of *Eleutherococcus senticosus* extracts on hexobarbital metabolism *in vivo* and *in vitro*. *J Ethnopharmacol* 10: 235-241.

SIRISA (INDIAN WALL NUT)

Botanical Name: *Albizzia lebbeck* (L.) Benth.

Natural Order: Fabaceae.

Active Ingredients: Saponins: albiziasaponins A, B, and C, alkaloids and tannin.

Non-Clinical Studies: Antianaphylactic (Tripathi and Das 1977; Barua *et al* 1997), anti-asthmatic (Tripathi and Das 1977), anticonvulsant (Kasture *et al* 2000), cytotoxic (Rahaman *et al*), nootropic (Chintwar *et al* 2002).

BIBLIOGRAPHY

Barua, C.C. *et al.* (1997). Studies on antianaphylactic activity of fractions of *Albizzia lebbeck*. *Cur Sci* 72: 397-9.

Chintwar, S.D. *et al.* (2002). Nootropic activity of *Albizzia lebbeck* in mice. *J Ethnopharmacology* 81 (3): 299-305.

Kasture, V.S. *et al.* (2000). Anticonvulsive activity of *Albizzia lebbeck, Hibiscus rosa sinesis* and *Butea monosperma* in experimental animals. *J Ethnopharmacology* 71 (1-2): 65-75.

Rahaman, N. *et al.* Anticancer agent from *Albizzia lebbeck*. ISHA Acta Horticulturae. V International Symposium on Medicinal, Aromatic and Spice Plants.

Tripathi, R.M. and Das, P.K. (1977). antianaphylactic and anti-asthmatic activity of *Albizzia lebbeck*. *Indian Journal of Pharmacology* 9 (3): 189-194.

SOLMON'S SEAL

Botanical Name: *Polygonum multiflorum* Thunb.

Natural Order: Polygonaceae

Active Constituents: Saponins, anthraquinones: emodin, physcion (Lin *et al* 2003), stilbene glucosides (Ryu *et al* 2002).

Non-Clinical Studies: Anti-ischemic (Chan, Cheng and Wang, 2003), antioxidant (Ryu *et al* 2002; Chan, Cheng and Wang, 2003), nootropic (Chan, Cheng and Wang 2002).

Adverse Drug Reactions: Contact dermatitis (Steyn 1934).

Toxicity: Hepatotoxic (Park *et al* 2001).

BIBLIOGRAPHY

Chan, Y.C.; Cheng, F.C. and Wang, M.F. (2002). Beneficial effects of different *Polygonum multiflorum* Thunb. extracts on memory and hippocampus morphology. *Journal of Nutritional Science and Vitaminology* (Tokyo). 48 (6): 491-497.

Chan, Y.C.; Cheng, F.C. and Wang, M.F. (2003). Long-term administration of *Polygonum multiflorum* Thunb. reduces cerebral ischemia-induced infarct volume in gerbils. *American Journal of Chinese Medicine* 31 (1): 71-77.

Chan, Y.C.; Cheng, F.C. and Wang, M.F. (2003). *Polygonum multiflorum* extracts improve cognitive performance in senescence accelerated mice. *American Journal of Chinese Medicine* 31 (2): 171-179.

Chen, Y. *et al.,* (1999). 2, 2-Diphenyl-1-picrylhydrazyl radical-scavenging active components from *Polygonum multiflorum* thunb. *Journal of Agricultural and Food Chemistry*. 47(6):2226-2228.

Lin, L.C. *et al.,* (2003). Micropropagation of *Polygonum multiflorum* Thunb and quantitative analysis of the anthraquinones emodin and physcion formed *In vitro* propagated shoots and plants. *Biology and Pharmacology Bulletin* 26 (10): 1467-1471.

Park, G.J. *et al.,* (2001). Acute hepatitis induced by Shou-Wu-Pian, an herbal product derived from *Polygonum multiflorum. Journal of Gastroenterology and Hepatology* 16 (1): 115-117.

Ryu, G. *et al.,* (2002). The radical scavenging effects of stilbene glucosides from *Polygonum multiflorum. Archives of Pharmacal Research* 25 (5): 636-639.

STONE ROOT

Botanical Name: *Collinsonia canadensis* Linn.

Natural Order: Lamiaceae.

Active Ingredients: Saponins (Joshi *et al* 1992).

BIBLIOGRAPHY

Joshi, B.S. *et al.* (1992). Saponins from *Callinsonia canadensis. J Nat Prod* 55 (10): 1468-1476.

SWEET BUGLEWEED

Botanical Name: *Lycopus virginicus* L

Natural Order: Lamiaceae.

Active Ingredients: Tannins, lithospermic acid, alkaloid: lycopine, phenolic derivatives: caffeic acid, chlorogenic acid, ellagic acid, rosmarinic acid, essential oil.

Non-Clinical Studies: Anti-thyrotropic (Winterhoff *et al* 1994) and antigonadotropic.

BIBLIOGRAPHY

Winterhoff, H. *et al.* (1994). Endocrine effects of *Lycopus eurapaeus* L. following oral application. *Arzneimittel forschung* 44: 41-45.

SWEET HERB

Botanical Name: *Stevia rebaudiana* Bert.

Natural Order: Asteraceae

Active Constituents: diterpene glycosides; stevioside (Bridel and Lavieille 1931; Haga, Ise and Kobayashi 1976; Ahmed and Dobberstein, 1982), steviolbioside (Ahmed, Dobberstein and Farnsworth, 1980; Ahmed and Dobberstein, 1982), dulcoside A (Kobayashi *et al* 1977; Ahmed and Dobberstein, 1982), dulcoside B (rebaudioside A (Kobayashi, *et al* 1977; Ahmed and Dobberstein, 1982), rebaudioside B (Ahmed, Dobberstein and Farnsworth, 1980; Ahmed and Dobberstein, 1982; Ahmed and Dobberstein, 1982), rebaudioside C (Ahmed and Dobberstein, 1982; Ahmed and Dobberstein, 1982), diterpenes: sterebins A-H (Oshima, Saito and Hikino, 1986), flavonoid glycosides, essential oil, phytosterols (Nabeta, Kasai and Sugisawa, 1976).

Non-Clinical Studies: Antihyperglycemic (Hübler Bracht and Kelmer-Bracht 1994; Jeppesen *et al* 2000; Jeppesen *et al* 2003). antihypertensive (Lee *et al* 2001; Liu *et al* 2003), contraceptive (Planas and Kuc 1968; Jeppesen, *et al* 2003).

Clinical Studies

Glucose tolerance test

Effect of water extract of *Stevia rebaudiana* on glucose tolerance was investigated in sixteen normal adult humans. Aqueous extracts of the leaves in a dose of 5 grams was administered to volunteers at regular 6 hour intervals for a time period of 3 days. Glucose tolerance tests were done prior to and after extract administration. A second group comprising six normal volunteers was instructed to consume an aqueous arabinose solution to study stress effects. The water extract of *Stevia rebaudiana* resulted in increase in glucose tolerance. The extract significantly decreased plasma glucose levels during the test and after overnight fasting in all volunteers (Curi *et al* 1986)

Hyperglycemia

In an acute cross-over study, effect of stevioside was studied in two diabetic patients. A standard test meal was supplemented with either 1 g of stevioside or 1 g of maize starch in the control group. Compared to control, stevioside reduced the incremental area under the glucose response curve by eighteen per cent. Further the insulinogenic index was increase by steviosides up to an extent of 40 per cent. The authors concluded that steviosides have positive effect on glucose metabolism (Gregersen, 2004).

Hypertension

A multicentre, randomized, double-blind, placebo-controlled study was conducted in China on one hundred six hypertensive patients (ages ranging from 28 to 75 years. Each subject was given capsules containing stevioside (250 mg) or placebo thrice daily and followed-up at monthly intervals for 1 year. After 3 months, the systolic and diastolic blood pressure of the stevioside group decreased significantly and the effect persisted during the whole year. No significant side effect was observed. (Chan, Tomlinson, Chen *et al* 2000).

Toxicity: LD50 of steviol: 6 g/kg to hamsters (Toskulkao, Chaturat, Temcharoen and Glinsukon 1997). LD50 of stevioside: 15 g/kg orally and 8.2 g/kg s.c. (Toskulkao, Chaturat, Temcharoen and Glinsukon, 1997; Xili, *et al* 1992).

LD 50 of aqueous extract of *Stevia rebaudiana* containing 50% steviosides: 3.4g/kg i.p. (Melis 1995), Stevioside and steviol lacks mutagenicity and genotoxicity (Klongpanichpak *et al* 1997; Matsui, *et al* 1996).

BIBLIOGRAPHY

Abudula R, *et al.,* (2004). Rebaudioside A potently stimulates insulin secretion from isolated mouse islets: Studies on the dose-, glucose- and calcium-dependency. *Metabolism* 53(10): 1378-1381.

Ahmed, M.S. and Dobberstein, R.H. (1982a). *Stevia rebaudiana.* II. High-performance liquid chromatographic separation and quantitation of stevioside, rebaudioside A and rebaudioside C. *J Chromatogr* 236: 523-526.

Ahmed, M.S. and Dobberstein, R.H. (1982b). *Stevia rebaudiana.* III. High-performance liquid chromatographic separation and quantitation of rebaudiosides B, D and E, dulcoside A and steviolbioside. *J Chromatogr* 245: 373-376.

Ahmed, M.S., Dobberstein, R.H. and Farnsworth, N.R. (1980). *Stevia rebaudiana.* I. Use of p-bromophenacyl bromide to enhance ultraviolet detection of water-soluble organic acids (steviolbioside and rebaudioside B) in high-performance liquid chromatographic analysis. *J Chromatogr* 192: 387-393.

Chan P., *et al.* (2000). A double-blind placebo-controlled study of the effectiveness and tolerability of oral stevioside in human hypertension. *Br J Clin Pharmacol* 50: 215–220.

Curi R. *et al.* (1986). Effect of *Stevia rebaudiana* on glucose tolerance in normal adult humans. *Brazilian Journal of Medical and Biological Research* 19 (6): 771-774.

Das, S., *et al.* (1992). Evaluation of the cariogenic potential of the intense natural sweeteners stevioside and rebaudioside A. *Caries Res* 26: 363-366.

Gregersen, S., *et al.* (2004). Antihyperglycemic effects of stevioside in type 2 diabetic subjects. *Metabolism* 53 (1): 73-76.

Hübler, M.O.; Bracht, A. and Kelmer-Bracht, A.M. (1994). Influence of stevioside on hepatic glycogen levels in fasted rats. *Research Communications in Chemistry, Pathology and Pharmacology* 84 (1): 111-118.

Jeppesen, P.B. *et al.* (2000). Stevioside acts directly on pancreatic beta cells to secrete insulin: actions independent of cyclic adenosine monophosphate and adenosine triphosphate-sensitive K+-channel activity. *Metabolism* 49 (2): 208-214.

Jeppesen, P.B. *et al.* (2003). Antihyperglycemic and blood pressure-reducing effects of stevioside in the diabetic Goto-Kakizaki rat. *Metabolism* 52 (3): 372-378.

Klongpanichpak, S. *et al.* (1997). Lack of mutagenicity of stevioside and steviol in *Salmonella typhimurium* TA 98 and TA 100. *Journal of the Medical Association of Thailand* 80 (1): 121-S128.

Kobayashi, M., *et al.* (1977). Dulcosides A and B, new diterpene glycosides from *Stevia rebaudiana. Phytochemistry* 16: 1405-1408.

Lee, C.N. *et al.* (2001). Inhibitory effect of stevioside on calcium influx to produce antihypertension. *Planta Medica.* 67(9):796-799.

Liu, J.C., *et al.* (2003). Mechanism of the antihypertensive effect of stevioside in anesthetized dogs. *Pharmacology* 67 (1): 14-20.

Matsui, M *et al.* (1996). Evaluation of the genotoxicity of stevioside and steviol using six *In vitro* and one *in vivo* mutagenicity assays. *Mutagenesis* 11: 573-579.

Melis, M.S. (1995). Chronic administration of aqueous extract of *Stevia rebaudiana* in rats: renal effects. *Journal of Ethnopharmacology* 47 (3): 129-134.

Nabeta, K., Kasai, T. and Sugisawa, H. (1976). Phytosterol from the callus of *Stevia rebaudiana. Agric Biol Chem* 40: 2103-2104.

Oliveira-Filho, R.M.; Uehara, O.A., Minetti, C.A. and Valle, L.B. (1989). Chronic administration of aqueous extract of *Stevia rebaudiana* (Bert.) Bertoni in rats: endocrine effects. *General Pharmacology* 20 (2): 187-191.

Oshima, Y., Saito, J. and Hikino, H. (1986). Sterebins A, B, C and D, bisnorditerpenoids of *Stevia rebaudiana* leaves. *Tetrahedron* 42: 6443-6446.

Planas, G.M. and Kuc, J. (1968). Contraceptive properties of *Stevia rebaudiana. Science* 162: 1007.

Toskulkao, C.; Chaturat, L.; Temcharoen, P. and Glinsukon, T. (1997). Acute toxicity of stevioside, a natural sweetener, and its metabolite, steviol, in several animal species. *Drug and Chemical Toxicology* 20 (1-2): 31-44.

Wasuntarawat, C.; Temcharoen, P.; Toskulkao, C.; Mungkornkarn, P. Suttajit, M. and Glinsukon, T. (1998). Developmental toxicity of steviol, a metabolite of stevioside, in the hamster. *Drug and Chemical Toxicology* 21 (2): 207-222.

Xili, L. *et al.* (1992). Chronic oral toxicity and carcinogenicity study of stevioside in rats. *Food Chem Tox* 30: 957-965.

TULSI

Botanical Name: *Ocimum sanctum* Linn.

Natural Order: Lamiaceae

Active Ingredients: Volatile oil: eugenol, methyl eugenol and caryophyllene (Dey and Choudhri 1980), triterpenoids: ursolic acid and rosmarinic acid, flavonoids: apigenin and luteolin and their glycosides, cirsilineol, cirsimaritin, isothymusin, isothy-monin, orientin molludistin and vicenin (Devi Bisht and Vinitha 1998), polyphenols (Skalsta *et al* 1999).

Non-Clinical Studies: Adaptogen (Bhargava and Singh 1981; Maity *et al* 2000), anticancer (Aruna and Sivaram Krishnan 1984; Banerjee *et al* 1996; Karthikeyan Ravichandran and Govindasamy 1999), anti-cataract (Adikary *et al* 1999), anti-fertility (Batta and Santha Kumari 1971), antioxidant (Kelm *et al* 2000; Shyamala and Devaki 1996), anti-inflammatory, analgesic and anti-pyretic (Godhwani Godhwani and Vyas 1987), anti-microbial (Dey and Choudhri 1984), anti-spermatogenic (Seth Johri and Sundaram 1981), antitubercular (Redi, Shukla and Singh 1986), anti-tussive (Nadiq and Lakshmi 2005), gastro protective (Mandal *et al* 1993; Singh, and Majumdar, 1999), cardio protective (Sood, *et al* 2005), hypoglycemic (Chattopadhyay 1993; Rai, Iyer and Mani 1997), immunomodulator (Mediratta *et al* 1988), and radio protective (Devi, Bisht and Vinitha 1998; Ganasoundari, Devi and Rao 1997; Ganasoundari, Devi and Rao 1998).

Clinical Studies: A randomised, placebo-controlled, single-blind, crossover trial in patients diagnosed with noninsulin-dependent diabetes mellitus was conducted in forty patients to study the effects of *Ocimum sanctum* (dried leaf 2.5 g daily) on fasting and postprandial blood glucose and serum lipid (cholesterol) levels. *Ocimum sanctum* caused a significant decrease in both fasting and postprandial blood glucose levels compared with placebo. A mild hypolipidemic action was reported (Agrawal, Rai and Singh, 1996).

Toxicity: LD_{50} in rats: ethanolic extract of the whole plant: 4508 ± 80 mg, oral (Khanna, Gupta and Grover 1986). LD 50 of the alcoholic extratct of the leaf: i.p., 3.75 mg.kg. It affects fertility (Seth, Johri and Sundaram 1981).

BIBLIOGRAPHY

Adikary, N. *et al.,* (1999). Studies on anti-cataract activity of *Ocimum sanctum* in experimental models of cataractogenesis. *Indian J. Pharamacol* 312 (1): 77.

Agarwal, P. *et al.* (1996): Randomized placebo-controlled, single blind trial of the holy basil leaves in patients with non-insulin-dependent diabetes mellitus. *Int. J Clin Pharmacol Ther* 34 (9): 406-409.

Agrawal, P., Rai, V. and Singh, R.B. (1996). Randomized placebo-controlled, single blind trial of holy basil leaves in patients with noninsulin-dependent diabetes mellitus. *International Journal of Clinical Pharmacology and Therapeutics* 34: 406-409.

Archana, R. and Namasivayam, A. (2000). Effect of *Ocimum sanctum* on noise induced changes in neutrophil functions. *Journal of Ethnopharmacology* 73: 81-85.

Aruna, K. and Sivaram Krishnan, V.M. (1984). Anticarcinogenic effects of the essential oils from cumin, poppyand basil. *Phytotherapy Res* 10 (7): 577-580.

Banerjee, S. *et al.* (1996). Modulatory influence of alcoholic extract of *Ocimum* leaves on carcinogen-metabolizing enzyme activities and reduced glutathione levels in mouse. *Nutrition and Cancer* 25 205-217.

Batta, S.K. and Santha Kumari, G. (1971). The antifertility effect of *Ocimum sanctum* and *Hibiscus rosa-sinensis*, *Indian J Med Res* 59 (5): 777-781.

Bhargava, K.P. and Singh, N. (1981). Anti-stress activity of *Ocimum sanctum* Linn. *Indian Journal of Medical Research* 73: 443-451.

Chattopadhyay, R.R. (1993). Hypoglycemic effect of *Ocimum sanctum* leaf extract in normal and streptozotocin diabetic rats. *Indian Journal of Experimental Biology* 31: 891-893.

Devi, P.U., Bisht, K.S. and Vinitha, M. (1998). A comparative study of radioprotection by *Ocimum* flavonoids and synthetic aminothiol protectors in the mouse. *British Journal of Radiology* 71: 782-784.

Dey, H.B. and Choudhri, M.A. (1980). Effect of plant development stage and some micronutrients on essential oil and eugenol content in *Ocimum sanctum* L. *Indian Perfumer* 24(4): 199-203.

Dey, H.B. and Choudhri, M.A. (1984). Essential oil of *Ocimum sanctum* L. and its antimicrobial activity. *Indian Perfumr* 28 (2): 82-87.

Ganasoundari, A., Devi, P.U. and Rao, B.S. (1998). Enhancement of bone marrow radioprotection and reduction of WR-2721 toxicity by *Ocimum sanctum*. *Mutation Research* 397: 303-312.

Ganasoundari, A., Devi, P.U. and Rao, M.N. (1997). Protection against radiation-induced chromosome damage in mouse bone marrow by *Ocimum sanctum*. *Mutation Research* 373: 271-276.

Godhwani, S., Godhwani, J.L. and Vyas, D.S. (1988). *Ocimum sanctum* - A preliminary study evaluating its immunoregulatory profile in albino rats. *Journal of Ethnopharmacology* 24: 193-198.

Godhwani, S., Godhwani, J.L. and Vyas, D.S. (1987). *Ocimum sanctum* : An experimental study evaluating its anti-inflammatory, analgesic and antipyretic activity in animals. *Journal of Ethnopharmacology* 21: 152-163.

Jangde, C.R. *et al.* (1996). Effect of *Ocimum sanctum* Linn. on bledding time in rats, *Int J Animal Sci* 11 (1): 249-250.

Karthikeyan, K., Ravichandran, P. and Govindasamy, S. (1999). Chemo-preventive effect of *Ocimum sanctum* on DMBA-induced hamster buccal pouch carcinogenesis. *Oral Oncology* 35: 112-119.

Kelm, M.A. *et al.* (2000). Antioxidant and cyclooxygenase inhibitory phenolic compounds from *Ocimum sanctum* Linn. *Phytomedicine* 7: 7-13.

Khanna, S., Gupta, S.R. and Grover, J.K. (1986). Effect of long term feeding of tulsi *(Ocimum sanctum* Linn.) on reproductive performance of adult albino rats. *Indian Journal of Experimental Biology* 24: 302-304.

Maity, T.K. *et al.* (2000). Effect of *Ocimum sanctum* roots extract on swimming performance in mice. *Phytotherapy Research* 14: 120-121.

Mandal, S. *et al.* (1993). *Ocimum sanctum* Linn - a study on gastric ulceration and gastric secretion in rats. *Indian Journal of Physiology and Pharmacology* 37: 91-92.

Mediratta, P.K. *et al.* (1988). Effect of *Ocimum sanctum* Linn. on humoral immune responses. *Indian Journal of Medical Research* 87: 384-386.

Nadiq, P. and Lakshmi, S. (2005). Study of anti-tussive activity of *Ocimum sanctum* Linn in guinea pigs. *Indian J Physiol Pharmacol* 49 (2): 243-5.

Prakash, J. and Gupta, S.K. (2000). Chemopreventive activity of *Ocimum sanctum* seed oil. *Journal of Ethnopharmacology* 72: 29-34.

Rai, V., Iyer, U. and Mani, U.V. (1997). Effect of Tulasi *(Ocimum sanctum)* leaf powder supplementation on blood sugar levels, serum lipids and tissue lipids in diabetic rats. *Plant Foods for Human Nutrition* 50: 9-16.

Redi, G.G., Shukla, N.P. and Singh, K.V. (1986). Chemotherapy of tuberculosis- antitubercular activity of *Ocimum sanctum* leaf extract, *Fitoterpia* 57 (2): 114-116.

Sakina, M.R. *et al.* (1990). Preliminary psychopharmacological evaluation of *Ocimum sanctum* leaf extract. *Journal of Ethnopharmacology* 28: 143-150.

Sembulingam, K., Sembulingam, P. and Namasivayam, A. (1997). Effect of *Ocimum sanctum* Linn on noise induced changes in plasma corticosterone level. *Indian Journal of Physiology and Pharmacology* 41: 139-143.

Seth, S.D., Johri, N. and Sundaram, K.R. (1981). Antispermatogenic effect of *Ocimum sanctum*. *Indian Journal of Experimental Biology* 19: 975-976.

Sharma, P., Kulshreshtha, S. and Sharma, A.L. (1998). Anti-cataract activity of *Ocimum sanctum* on experimental cataract. *Indian Journal of Pharmacology* 30: 16-20.

Shyamala, A.C. and Devaki, T. (1996). Studies on peroxidation in rats ingesting copper sulphate and effect of subsequent treatment with *Ocimum sanctum*. *Journal of Clinical Biochemistry and Nutrition* 20: 113-119.

Skalsta, H, *et al.* (1999). Polyphenols of *Ocimum sanctum* from Surinam. *Fitoterapia* 37 (1): 92-94.

Singh, S., Majumdar, D.K. and Rehan, H.M. (1996). Evaluation of anti-inflammatory potential of fixed oil of *Ocimum sanctum* (Holybasil) and its possible mechanism of action. *Journal of Ethnopharmacology* 54: 19-26.

Singh, S. and Majumdar, D.K. (1997). Evaluation of anti inflammatory activity of fatty acids of *Ocimum sanctum* fixed oil. *Indian Journal of Experimental Biology* 35: 380-383.

Singh, S. and Majumdar, D.K. (1999). Evaluation of the gastric antiulcer activity of fixed oil of *Ocimum sanctum* (Holy Basil). *Journal of Ethnopharmacology* 65: 13-19.

Singh, S. (1998). Comparative evaluation of antiinflammatory potential of fixed oil of different species of *Ocimum* and its possible mechanism of action. *Indian Journal of Experimental Biology* 36: 1028-1031.

Sood, S *et al.,* (2005). Chronic oral administration of *Ocimum sanctum* Linn. augments cardiac endogenous antioxidants and prevents isoproterenol-induced myocardial necrosis in rats. *J Pharm Pharmacol* 57 (1): 127-33.

TUVRAKA (JUNGLE ALMOND)

Botanical Name: *Hydnocarpus wightiana* Blume.

Natural Order: Flacourtiaceae

Active Ingredients: Fixed oil containing chaulmoogric acid and hydnocarpic acid, flavonolignans: hydnowightin, hydnocarpin, neohydnocarpin (Sharma and Hall 1991).

Non-Clinical Studies: Hypolipidemic, anti-inflammatory, antineoplastic (Sharma and Hall 1991).

BIBLIOGRAPHY

Sharma, D.K. and Hall, I.H. (1991). Hypolipidemic, anti-inflammatory and antineoplastic activity and cytotoxicity of flavonolignans isolated from *Hydnocarpus wightiana* seeds. *Nat Prod* 54 (5): 1298-302.

VARUNA (THREE –LEAVED CAPER)

Botanical Name: *Cratavea nurvala* Buch-Ham. Syn. *Cratavea religiosa* Hook.f. and Thomas.

Natural Order: Papaveraceae

Active Ingredients: Saponin, tannins, triterpenoids: lupeol, varunol, choline.

Non-Clinical Studies: Anti-inflammatory (Geetha and Varalakshmi, 1999). antiseptic, anti-utolithic (Varalakshmi, Shamila and Latha 1990) and nephroprotective (Shirwakar *et al* 2004).

BIBLIOGRAPHY

Geetha, T and Varalakshmi, P. (1999). Anticomplement activity of triterpenes from *Crataeva nurvala* stem bark in adjuvant arthritis in rats. *Gen. Pharmacol* 32 (4): 495-7.

Shirwakar, A. *et al.* (2004). Effect of lupeol isolated from *Crataeva nurvala* Buch.-Ham. stem bark extract against free radical induced nephrotoxicity in rats. *Indian J Exp Biol* 42: 686-690.

Varalakshmi, P.; Shamila, Y. and Latha, E. (1990). Effect of *Crataeva nurvala* in experimental urolithiasis. *J Ethnopharmacol* 28(3): 313-321.

VASAKA

Botanical Name: *Adhatoda vasica* Nees.

Natural Order: Acanthaceae

Active Ingredients: Alkaloids; Adhatodine, adhavasinone (Abd and Ahmed 1998), anisotine (Abd and Ahmed 1998), 6-hydroxy-peganine, deoxyvasicinone (Abd and Ahmed 1998). deoxyvasicine, vasicine or peganine (De and Ray, 1927; Morris, Handford and Adama, 1935; Glasby 1978, Gupta *et al* 1978; Bhalla *et al* 1982; Chandoke, 1987; Arambewela *et al* 1988; Rajani and Pundarikakshudu 1996; Mhaske and Argade 2001: Das, Poi and Chawdhury, 2005), vasicol (Dhar *et al* 1981), (+)–vasicinone, it is auto-oxidation product of vasicine (Bhalla *et al* 1982; Lakshmi, Poi and Chaudhary 1988; Mhaske and Argade 2001; Das, Poi and Chawdhury 2005), vasicinolone, vasicoline, 9-acetamido-3,4-dihydropyrido [3,4-b] indole, 6, 7, 8, 9, 10, 12-hexahydro-azepino-(2, 1-b) quinazolin-12-one-(Johri and Zutshi 2000), aliphatic hydroxyketones (Singh, *et al* 1991), aliphatic alcohol (Singh *et al* 1991), flavonoids: quercetin (Rawat, *et al* 1994; Ahmed *et al* 1999), glucoside: 2'-hydroxy-4-glucosyl-oxychalcone, O-ethyl-a -D-galactoside and non-nitrogenous principle: vasakin.

Non-Clinical Studies: Anti-anaphylactic, anti-diabetic, anti bacterial (Chopra 1925; Patel and Venkatakrishna-Bhatt, 1984). anti-inflammatory (Chakraborty and Brantner 2001), abortifacient (Atal 1980), anti-tubercular (Gupta and Chopra, 1954), antioxidant (Singh and Padmavathi, 2000), anti-tussive (Dhuley, 1999). bronchodilator (Amih and Mehta 1959; Johri and Zutshi 2000). hepatoprotective (Bhattacharya *et al* 2005). hypotensive, thrombopoietic (Sharma and Atal 1985; Atal *et al* 1982), oxytocic (Atal, 1980; Gautam and Sharma 1981; Sharma and Atal 1985; Zutshi *et al* 1980; Maddapa, Sankaranarayan and Sharma 1989; Lal and Sharma, 1981) and radio protective (Kumar *et al* 2005)

Toxicity: Abortifacient (Nath *et al* 1992). A study was carried out to determine the acute toxicity of vasicine in humans. 24 human volunteers with 0.5-16 mg dose of vasicine injected intravenous (in 500 ml saline in 3 hours). Vasicine tried up to 16 mg dose on the hospital in-patients on 2nd to 8th day of normal puerperium was well tolerated and showed no undesirable effect in clinical observations, haematological and biochemical investigations and kidney and liver function tests carried out before, during and after vasicine treatment. However, uterus became firm and contracted after vasicine treatment which indicated its effectiveness as an oxytocic (Wakhloo *et al* 1980). Vasicine is non-teratogenic.

BIBLIOGRAPHY

Abd, E.M.H.F. and Ahmed, E.S. (1998). Isoquinoline and quinazoline alkaloids of *Adhatoda vasica* Nees *Pharmaceutical and Pharmacological Letters* 8 (4): 167-69.

Ahmed, E.S.S. *et al.* (1999). Flavonoids and antimicrobial volatiles from *Adhatoda vasica* Nees. *Pharmaceutical and Pharmacological Letters* 9 (2): 52-56.

Amih, A.H. and Mehta, D.R. (1959). A bronchodilator alkaloid (vasicinone) from *Adhatoda vasica* Nees. *Nature* 184 (17): 1317.

Arambewela, A.S. *et al.* (1988). Vasicine contents and their seasonal variation in *Adathoda vasica, Fitoterapia* 59 (2): 151-153.

Atal, C.K. (1980). Chemistry and Pharmacology of Vasicine: A New Oxytoxic and Abortifacient. RRL, Jammu, India.

Atal, C.K. *et al.* (1982). Thrombopoietic activity of vasicine hydrochloride. *Ind J Exp Biol* 20(9): 704-9.

Bhalla, H.L. *et al.* (1982). Improved method extraction and analysis of vasicine and vasicinone. The alkaloids of *Adathoda vasica* Nees. *Indian drugs* 20:16.

Bhattacharaya, D. *et al.,* (2005). Hepatoprotective activity of *Adhatoda vasica* aqueous leaf extract on D-galactosamine-induced liver damage in rats. *Fitoterapia* 76 (2): 223-5.

Chakraborty, A and Brantner, A.H. (2001). Study of alkaloids from *Adhatoda vasica* Nees on their anti-inflammatory activity. *Phytother Res* 15 (6): 532-4.

Chandoke, N. (1987). Vasicine, the alkaloid of *Adathoda vasica* a novel abortifacient, *Indian Drugs* 24 (9): 425-429.

Chopra, R.N. (1925). Some observations on the pharmacological action and therapeutic properties of *Adhatoda vasica. Ind J Med Res* 13: 205.

Chowdhury, B.K., Hirani, S.K. and Ngur, D. (1987). High-performance liquid chromatographic study of the photochemical oxidation of vasicine and its analogues. *J Chromatogr* 390 (2):439-43.

Das, C., Poi, R. and Chawdhury, A. (2005). HPTLC determination of vasicine and vasicinone in *Adhatoda vasica. Phytochem Anal* 16(2): 90-2.

Dhar, K.L. *et al.* (1981). Vasicol, A new alkaloid from *Adathoda vasica, Phytochemistry* 20 319-321.

De, A.K. and Ray, J.N. (1927). The constitution of vasicine, *Jour Indian Chem Soc* 4: 541-545.

Dhuley, J.N. (1999). Antitussive effect of *Adhatoda vasica* extract on mechanical or chemical stimulation-induced coughing in animals. *J Ethnopharmacol* 67 (3): 361-5.

Gautam, C.S. and Sharma, P.L. (1981). Potentiation of oxytocin evoked responses by (+) sotalol deoxysotalol and Vasicine HCl on isolated rat and rabbit uterus. *Indian J Med Res* 76 107-14.

Glasby, J.S. (1978). Encyclopedia of the alkaloids. London: Plenum Press p 1367.

Gupta, K.C. and Chopra, I.C. (1954). Anti-tubercular action of *Adathoda vasica* (N.O. Acanthaceae). *Indian J Med Res* 42 (3): 135.

Gupta, O.P. *et al.* (1978). Vasicine, alkaloid of *Adathoda vasica,*a promising uterotonic abortifacient. *Indian J Exp Biol* 16 (10): 1075-1077.

Gupta, O.P. *et al.* (1978). Sub acute toxicity and teratogenic studies on vasicine - the alkaloid of *Adathoda vasica*: A new potent uterine stimulant. *Indian J Pharmac* 10 (1): 87.

Johne, S. *et al.* (1973). Studies on biosynthesis of the quinazoline alkaloid peganin in *Adathoda vasica* Nees. *Pharmazie* 28 (6): 403-406.

Johri, R. K. and U. Zutshi (2000). Mechanism of action of 6, 7, 8, 9, 10, 12-hexahydro-azepino-(2, 1-b) quinazolin-12-one-(RLX): A novel bronchodilator. *Indian Journal of Physiology and Pharmacology* 44 (1): 75-81.

Kumar, A. *et al.* (2005). Modulatory influence of *Adhatoda vasica* Nees leaf extract against gamma irradiation in Swiss albino mice. *Phytomedicine* 12 (4): 285-93.

Lakshmi Poi, R.A. and Aditya Chaudhary, N. (1988). Occurrence of (+)– vasicinone in *Adathoda vasica Nees., J Ind Chem Soc* 65(11) 814.

Lal, R. and Sharma, P.L. (1981). Potentiation of prostaglandin evoked contractions of isolated rat uterus by vasicine hydrochloride. *Indian J Med Res* 73: 641-8.

Maddapa, C., Sankaranarayan, A. and Sharma, P.L. (1989). A study on the selectivity of action of (+) INPEA and vasicine in different isolated tissue preparations. *Indian Journal of Pharmacology* 21 (4): 144-152.

Mhaske, S.B. and Argade, N.P. (2001). Concise and efficient synthesis of bioactive natural products peganine, deoxyvasicinone, and (-)-vasicinone. *J Org Chem* 66 (26): 9038-40.

Morris, R.C.; Handford, W.E. and Adama, R. (1935). Structure of Vasicine, *J Amer Chem Soc* 57: 951-954.

Nath, D. *et al.* (1992). Commonly used Indian abortifacient plants with special reference to their teratologic effects in rats. *Journal of Ethnopharmacology* 36 (2): 147 154.

Pahwa, G.S *et al.* (1987). Chronic toxicity studies with vasicine from *Adathoda vasica Nees.*in rats and monkeys, *Indian J Exp Biol* 25 (7): 467-470.

Patel, P.K. and Venkatakrishna-Bhatt, H. (1984). *In vitro* study of antimicrobial activity of *Adhatoda vasica* (leaf extract) on gingival inflammation a preliminary report. *Ind J Med Sci* 38 (4): 70-2.

Rajani, M. and Pundarikakshudu, K. (1996). A note on the seasonal variation of alkaloids in Adhatoda vasica Nees. *International Journal of Pharmacognosy* 34 (4): 308-309.

Rajan, M. and Bhavasar, G.C. (1993). *Adathoda vasica Nees.,* Effect of Chloramphenicol on total alkaloids, *Indian Drugs* 30 (2): 87-88.

Rawat, M.S.M. *et al.* (1994). Biochemical investigation of some wild fruits of Garhawal Himalayas. *Prog Horticult* 26 (1-2): 35-40.

Sharma, M.L. and Atal, C.K. (1985). Oxytocic, Thrombopoietic and Bronchodilatory activites of vasicine - a novel molecule isolated from *Adathoda vasica,* Paper No. 4.03. Paper presented at workshop on selected Medicinal Plants in Traditional/Indigenous Systems of Medicine at Bombay

Singh R.P and Padmavathi (2000). Modulatory influence of *Adhatoda vasica* leaf extract on the Enzymes of xenobiotic metabolism, antioxidant status and lipid per oxidation in mice. *Mol. Cell Biochem* 213 (1-2): 99-109.

Singh, R.S. *et al.* (1991). Aliphatic hydroxyketones from *Adathoda vasica, Phytochemistry* 30 (11): 3799-3801.

Singh, R.S. *et al.* (1991). New Aliphatic alcohol from *Adathoda vasica, Phytochemistr* 63 (3): 262-263.

Thappa, R.K. *et al.,* (1996). Two pyrroloquinazolines from *Adhatoda vasica. Phytochemistry* 42 (5): 1485-1488.

Wakhloo, R.L. *et al.,* (1980). Safety of vasicine hydrochloride in human volunteers. *Indian Journal of Pharmacology* 12(2): 129-131.

Zutshi, U., *et al.,* (1980). Absorption and distribution of vasicine, a modern uterotonic. *Planta Med* (40): 373-377.

VIBHITUKA

Botanical Name: *Terminalia bellerica* Roxb.

Natural Order: Combretaceae

Active Ingredients: Ellagic acid, gallic acid (Anand, K.K. *et al.* 1997), chebulagic acid (Iyengar and Dwivedi 1989), ß-sitosterol and bellericanin (Awasthi and Nath 1968), triterpenoids and their glycosides (Nandy *et al* 1989). lignans: termilignan, thannilignan 7-hydroxy-3',4'-(methylenedioxy) flavan and anolignan B (Valsaraj *et al* 1996/97).

Pre-Clinical Studies: Antimalarial (Valsaraj *et al* 1996/97). antimicrobial (Nandy *et al* 1997), anti-mutagenic (Padam *et al* 1996). antifungal (Valsaraj *et al* 1996/97), cardio protective (Tariq *et al* 1977; Shaila *et al* 1995).

Clinical Studies

Cough and Asthma

Ninety-three cases of different age groups were selected for clinical testing of *Terminalia bellerica*. Sixty-one cases were suffering from cough, twelve form breathlessness and twenty from both (cough and breathlessness). *Terminalia bellerica* was given in a dose of 2-6 G thrice-a-day. Twenty-two cases demonstrated complete relief whereas thirty-five cases showed slight improvement (Trivedi, Nesamany and Sharma 1982).

BIBLIOGRAPHY

Ali, M. (1992). Chemical examination of fruits of *Terminalia bellirica* Roxb. *Oriental Chem* 8 (3) 255-256.

Anand, K.K., *et al.* (1994). Hepatoprotective studies of a fraction from the fruits of *Terminalia bellirica* Roxb on experimental liver injury in rodents. *Phytotherapy Res* 8(5): 287-292.

Anand, K.K. *et al.* (1997). 3, 4, 5-Trihyrroxy benzoic acid (gallic acid). the hepatoprotective principle in the fruits of *Terminalia bellirica*-bioassay guided activity. *Pharamacol Res* 36 (4): 315-323.

Awasthi, K.I. and Nath, B. (1968). Chemical examination of *Terminalia bellirica*, I, Cardiac glycoside. *Indian Chem Soc* 45 (10): 913-917.

Iyengar, M.A. and Dwivedi, S. (1989). *Terminalia bellerica* - (Bahera) - a review. *Indian Drugs* 26 (12): 635-663.

Merhra, P.N. and Dadry, S.M.J.S. (1963). Pharmacognostic study of fruit of *Terminalia bellerica* Roxb. *Planta Medica* 11 (2): 176-182.

Nandy, A.K. *et al.* (1989). Triterpenoids and their glycosides from *Terminalia bellerica*. *Phytochemistry* 28 (10) 2769-2772.

Nandy, A.K. *et al.* (1997). Antimicrobial activity of *Terminalia bellerica* triterpenoids. *Fitoterapia* 68 (2): 178-180.

Padam, S.K. *et al.* (1996). Antimutagenic effects of polyphenols from *Terminalia bellerica* myrobalan in *Salmonella typhimurium*. *Indian J Exp Biol* 34 (2): 98-102.

Shaila, H.P. *et al.* (1995). Preventive actions of *Terminalia bellerica* in experimentally induced atherosclerosis. *Int J Cardiol* 49 (2): 101-6.

Tariq, M. *et al.* (1977). Protective effect of fruit extracts of *Emblica officinalis* (Gaertn). and *Terminalia belerica* (Roxb.) in experimental myocardial necrosis in rats. *Indian J Exp Biol* 15 (6): 485-6.

Trivedi, V.P., Nesamany, S. and Sharma, V.K. (1982). A clinical study of the antitussive and antiasthmatic effect of *Vibhitakphal churana* (*Terminalia bellerica* Roxb.) in the cases of Kasa-swasa, *J Res Ayur Siddha* 3 (1-2) 1-8.

Valsaraj, R., *et al.*, (1996/97). Novel lignans from *Terminalia bellerica* with antimicrobial activity. *Phytomedicine* 3 (1): 81.

Valsaraj, R., *et al.*, (1997). New anti-HIV-1, antimalarial and antifungal compounds from *Terminalia bellerica. J Nat Prod* 60 (7): 739-742.

VRIKSHAMLA (GARCINIA)

Botanical Name: *Garcinia indica* Choisy. Syn. *Garcinia purpurea* Roxb.

Natural Order: Clusiaceae.

Active Ingredients: Bioflavonoids: morelloflavone, volkensiflavone (Koshy and Vijayalakshmi 2001), euxanthone, polyisoprenylated phenolic compounds or pigments; garcinol, isogarcinol, cambogin, camboginol (Cotterill, Scheinmann and Puranik 1977; Krishnamurthy, Lewis and Ravindarnath 1981; Krishnamurthy, and Ravindranath 1982; Koshy and Vijayalakshmi 2001), (-) hydroxycitric acid (Jena *et al* 2002), benzophenones and L- leucine (Badami and Razdan 1972).

Non-Clinical Studies: Appetite suppressant (Ohia *et al* 2002). anti-obesity (Sullivan *et al* 1972; Lowenstein 1971; Greenwood *et al* 1981; Hasegawa 2002), antioxidant (Mahendran and Shyamala Devi, 2001), central nervous system depressant, chemopreventive (Tanaka *et al* 2000), gastroprotective (Mahendran and Shyamala Devi 2001; Mahendran and Devi 2002), hypoglycermic (Hayamizu *et al.,* 2003), hypolipidemic (Mahendran and Devi 2001).

Clinical Studies: In a twelve-week randomized, double-blind, randomized controlled trial *Garcinia cambogia* failed to produce significant weight loss and fat mass loss beyond that observed with placebo (Heymsfield *et al* 1998).

Pharmacokinetics: Hydroxy-citric acid is predominantly used in salt form because of increased absorption. Further it increases stability of hydroxyl citric acid. Research finding however suggest that calcium and magnesium salts of (-)-hydroxycitric acid are slightly soluble in water and poorly absorbed in the gastrointestinal tract (Clouatre and Dunn, 2002).

Calcium and magnesium salts of (-)-hydroxycitric acid are acted upon by bile acids and lipids in the gut. They bound to fibers or pectin or secreted during digestion (Clouatre and Dunn, 2002).

The peak plasma level of (-)-hydroxycitric acid was 8.4: g/ml at two hours after oral administration of 2 g of *Garcinia cambogia* extract to normal subjects, suggesting a limited efficiency of (-)-hydroxycitric acid absorption. (-)-Hydroxycitric acid was also detected in the urine (Loe *et al* 2001).

Drug-Intractions: Antiarrhythmics, antihistaminics, beta-blockers, calcium-channel blockers, digoxin and muscle relaxants.

Adverse Drug Reactions: Ocular irritant in rabbits (Ohio*et al* 2002). More studies are required (Pittler Schimdt and Ernst 2005).

Contraindications: Pregnancy.

Toxicity

Acute toxicity (calcium/potassium salt)

LD 50 in rats > 5000 {gavage} LD 50 in rabbits > 2000 {dermal} (Ohio *et al* 2002).

Chronic toxicity: An animal study reported dose- and time-dependent effects of a (-)-hydroxycitric acid extract. The extract does not cause any changes in hepatic and testicular lipid peroxidation, DNA fragmentation, or histopathological changes (Shara *et al* 2003).

BIBLIOGRAPHY

Badami, R.C. and Razdan, M.K. (1972). Isolation and identification of L-leucine as DNP-L-leucine hydrochloride in the leaves of *Garcinia indica*, *J Chem Soc* 49: 583.

Clouatre, D.L. and Dunn, J.M. (2002). Potassium (-)-hydroxycitric acid methods for pharmaceutical preparations for stable and controlled delivery. US Patent # 6, 447, 807

Cotterill, P.J.; Scheinmann, F. and Puranik, G.S. (1977). Phenolic compounds from the heartwood of *Garcinia indica*, *Phytochemistry* 16 148-149.

Greenwood, M.R.C. *et al* (1981). Effect of (-)-hydroxycitrate on development of obesity in the Zucker obese rat. *Am J Physiol* 240:E72–8

Hasegawa, N. (2002): Garcinia extract inhibits lipid droplet accumulation without affecting adipose conversion in 3T3-L1 cells. *Phytotherapy Research* 15 (2): 172-173.

Hayamizu, K. *et al.,* (2003). Effect of *Garcinia cambogia* extract on serum leptin and insulin in mice. *Fitoterapia* 74 (3): 267-273.

Heymsfield, S.B. *et al.,* (1998): *Garcinia cambogia* (hydroxycitric acid) as a potential antiobesity agent: a randomized controlled trial. *Journal of the American Medical Association* 280 (18): 1596-1600.

Ishihara, K. *et al.,* (2000). Chronic (-)-hydroxycitrate administration spares carbohydrate utilization and promotes lipid oxidation during exercise in mice. *Journal of Nutrition*130 (12): 2990-2995.

Jena, B.S. *et al.,* (2002). Chemistry and biochemistry of (-)-hydroxycitric acid from Garcinia. *Journal of Agriculture and Food Chemistry* 50 (1): 10-22.

Koshy, A.S. and Vijayalakshmi, N.R. (2001). Impact of certain flavonoids on lipid profiles--potential action of *Garcinia cambogia* flavonoids. *Phytotherapy Research* 15 (5): 395-400.

Krishnamurthy, N. and Ravindranath, B. (1982). Crystal and molecular structure of isogarcinol, *Tetraherdron Letters* 23: 2233.

Krishnamurthy, N.; Lewis, Y.S. and Ravindarnath, B. (1981). On the structure of garcinol, isogarcinol and camboginol, *Tetraherdron Letters* 22 (8): 793-796.

Loe, Y.C., Bergeron, N., Rodriguez, N. and Schwarz, J.M. (2001). Gas chromatography/mass spectrometry method to quantify blood hydroxycitrate concentration. *Analytical Biochemistry* 292: 148-154.

Lowenstein, J.M. (1971): Effect of (-)-hydroxycitrate on fatty acid synthesis by rat liver *in vivo*. *J Biol Chem*; 246:629–32.

Mahendran, P. and Devi, C.S. (2001). Effect of *Garcinia cambogia* extract on lipids and lipoprotein composition in dexamethasone administered rats. *Indian Journal of Physiology and Pharmacology* 45 (3): 345-350.

Mahendran, P. and Shyamala Devi, C.S. (2001). The modulating effect of *Garcinia cambogia* extract on ethanol-induced peroxidative damage in rats. *Indian Journal of Pharmacology* 33: 87-91

Mahendran, P., Sabitha, K.E. and Devi, C.S. (2002): Prevention of HCl-ethanol induced gastric mucosal injury in rats by *Garcinia cambogia* extract and its possible mechanism of action. *Indian Journal of Experimental Biology*. 2002; 40(1):58-62.

Ohia, S.E. *et al.*, (2002): Safety and mechanism of appetite suppression by a novel hydroxycitric acid extract (HCA-SX). *Molecular and Cellular Biochemistry* 238 (1-2):89-103.

Pittler, M.H., Schimdt, K. and Ernst, E. (2005). Adverse events of herbal food supplements for body weight reduction: systematic review. *Obse Rev* 6 (2): 89-92.

Rao, A.V.R. and Venkatswamy, G. (1981). Camboginol, Garcinol and Isogarcinol, *Indian J Chem* 20B 983.

Shara, M. *et al.* (2003). Dose- and time-dependent effects of a novel (-)-hydroxycitric acid extract on body weight, hepatic and testicular lipid peroxidation, DNA fragmentation and histopathological data over a period of 90 days. *Mol Cell Biochem* 254 (1-2): 339-46.

Sullivan, A.C. *et al.* (1972). Inhibition of lipogenesis in rat liver by (-)-hydroxycitrate. *Arch Biochem Biophys* 150: 183–90.

Tanaka, T. *et al.*, (2000). Prevention of colonic aberrant crypt foci by dietary feeding of garcinol in male F344 rats. *Carcinogenesis* 21(6): 1183-1189,

WHITE BYRON

Botanical Name: *Bryonia alba* Linn.

Natural Order: Cucurbitaceae

Active Constituents: Cucurbitacins, triterpenes, lectins, sterols and glucoside: bryonin or bryonoside (Pohlmann 1980; Oobayashi, Yoshikawa and Arihara 1992).

Non-Clinical Studies: Anti-tumor and hypoglycemic (Anon: PDR for Herbal Medicines).

Toxicity: Toxic due to presence of cucurbitacins.

BIBLIOGRAPHY

Anon (2000). PDR for Herbal Medicines. Medical Economics Company Inc, Montvale, NJ: 812-813.

Oobayashi, K.; Yoshikawa, K. and Arihara, S. (1992). Structural revision of Bryonoside and structure elucidation of minoe saponins from *Bryonia dioca*. *PH* 31: 943-946.

Pohlmann, J. (1980). The cucurbitacins in *Bryonia alba* and *Bryonia dioica*. *PH* 14 (7): 1587-1589.

WILD DAISY

Botanical Name: *Bellis perennis* Linn.

Natural Order: Asteraceae

Active Constituents: Triterpene saponins (Schopke *et al* 1992). Flavonoids and polyacetylenes (Avato *et al* 1997).

Non-Clinical Studies: Antifungal (Willigmann *et al* 1993). antimicrobial (Avato *et al* 1997).

BIBLIOGRAPHY

Anon: (2000). PDR for Herbal Medicines. Medical Economics Company Inc, Montvale, NJ 812-813.

Avato, P. *et al.* (1997). Antimicrobial activity of polyacetylenes from *Bellis perennis* and their synthetic derivatives. *Planta Med* 63 (6): 503-7.

Schopke, T.H. *et al.* (1992). Triterpenoid saponins of plants of the Asteraceae tribe (Asteraceae). *Planta Med* 59 (7): A 591.

Willigmann, I. *et al.* (1993). Antimycotic compounds from different *Bellis perennis* varieties. *Planta Med* 61.

WILD INDIGO

Botanical Name: *Baptista tinctoria* Linn. ssp. *foetida* (Vis.) Hayek

Natural Order: Fabaceae

Active Constituents: Alkaloids: anagyrine, cytisine, sparteine (Anon: PDR for Herbal Medicines), isoflavonoid: formononetin (Anon: PDR for Herbal Medicines), hydroxycoumarin (Anon: PDR for Herbal Medicines) and glycoprotein.

Non-Clinical Studies: Immunomodulator (Beuscher and Kopanski 1985; Egert and Beuscher 1992) and estrogenic.

BIBLIOGRAPHY

Anon: (2000). PDR for Herbal Medicines. Medical Economics Company Inc, Montvale NJ; 812-813.

Bodinet, C.; Beuscher, N. and Kopanski, L. (2001). Purification of Immunological active glycoproteins from *Baptisia tinctoria* roots by affinity chromatography and isoeletric focusing. In: PM, Abstracts of the 37th Annual Congress on Medicinal Plant Research Braunschwe.

Egert, D. and Beuscher, N. (1992). Studies on antigen specifity of immunoreactive arabinogalactan proteins extracted from *Baptisia tinctoria* and *Echinacea purpurea. Planta Med* 58 (2): 163-5.

Beuscher, N. and Kopanski, L. (1985). Stimulation of immunity by the contents of *Baptisia tinctoria. Planta Med* 5: 381–4.

WILLOW HERB

Botanical Name: *Epilobium angustifolium* Linn. syn. *Epilobium parviflorum,* syn. *Epibolium hirsutum* Linn.

Natural Order: Onagariaceae.

Active Ingredients: Sterols, flavonoids, tannins: oenthein A and B.

Non-Clinical Studies: Anti-inflammatory, antimicrobial, anticancer (Durcey *et al* 1997).

BIBLIOGRAPHY

Durcey, B. *et al.* (1997). Inhibition of 5 alpha-reductase and aromatase by ellagitannins oenothein A and B from Epilobium species. *Planta Med* 63 (2): 111-114.

YARROW

Botanical Name: *Achillea millefolium* Linn. Syn *Achillea lanulosa* Nutt.

Natural Order: Asteraceae.

Active Constituents: Sesquiterpene lactones: 2, 3-dihydro-desacetoxy-matricin, 3 beta-hydroxy-11alpha, 13-dihydro-costunolide, 8, alpha-angeloyloxy-10-epi-artabsin, achillicin, 8-alphaangeloxy-leucodin, achimillic acids A, B and C, sterol: ß-sitosterol, flavonoids: apigenin, luteolin, rutin, centaureidin (Chandler, Hooper and Harvey, 1982; Ochir, Budesinky and Motl 1991: Tozyo *et al* 1994; Glasl *et al* 2002), flavones: 5-hydroxy- 3,6,7,4'-tetramethoxyflavone, artemetin, casticin, polyene: ponticaepoxide, alkaloid: achilleine or betonicine (Miller and Chow 1954). volatile oil: contains α- and ß-pinenes, borneol, bornyl acetate, camphor, caryophyllene, chamazulene eugenol, farnesene, guajazulene, myrcene, sabinene, salicylic acid, terpineol, thujone.

Non-Clinical Studies: Anti-inflammatory (Kastner *et al* 1993). diaphoretic (Chandler, Hooper and Harvey 1982), antispermatogenic (Montanari de Carvalho and Dolder 1998; Dalstner *et al* 2004), haemostatic.

Drug Interactions: Antineoplastic drugs.

Adverse Drug Reations: Dermatitis (Stingeni *et al* 1999).

Toxicity: Poisonous (due to presence of thujone) and hepatotoxic (due to presence of pyrrozolidine alkaloids).

BIBLIOGRAPHY

Chandler, R.F., S.N. Hooper and M.J. Harvey (1982). Ethnobotany and phytochemistry of yarrow, *Achillea millefolium*, Compositae. *Econ Bot* 36: 203-223.

Dalstner, P.R. *et al.* (2004). Reproductive evaluation of aqueous crude extract of *Achillea millefolium* L. (Asteraceae) in Wistar rats. *Reprod toxicol* 18 (6): 819-23.

Glasl, S. *et al.* (2002). Sesquiterpenes and flavonoid aglycones from a Hungarian taxon of the *Achillea millefolium* group. *Z Naturoforsch* 57 (11-12): 976-82.

Kastner, U *et al.* (1993). Anti-edematous activity of sesquiterpene lactones from different taxa of the *Achillea millefolium* group. *Planta Med* 59 (7): A 669.

Miller, F.M. and Chow, L.M. (1954). Alkaloids of *Achillea millefolium* L. I. Isolation and characterization of achilleine. *Journal of the American Chemical Society* 76(5): 1353-1354.

Montanari, T.; de Carvalho, J.E. and Dolder, H. (1998). Antispermatogenic effect of *Achillea millefolium* L. in mice. *Contraception* 58(5): 309-13.

Muller-Jakic, B. *et al.,* (1994). *In vitro* inhibition of cyclooxygenase and 5-lipoxygenase by alkamides from Echinacea and Achillea species. *Planta Med* 60:37.

Ochir, G.; Budesinky, M. and Motl, O (1991). M 3-Oxa-guaianolides from *Achillea millefolium. PH* 30 (12): 4163.

Stingeni, L. *et al.,* (1999). T-lymphocyte cytokine profiles in compositae airborne dermatitis. *Br J Dermatol* 141 (4): 689-93

Tozyo, T. *et al.,* (1994). Novel antitumor sesquiterpenoids in *Achillea millefolium. Chem Pharm Bull* (Tokyo) 42 (5):1096-100.

YASTIMADHU

Botanical Name: *Glycyrrhiza glabra* Linn.

Natural Order: Fabaceae.

Active Ingredients: Triterpenoid saponins; glycyrrhizin or glycyrrhizic acid its aglycone; glycyrrhetinic acid*, 18 ß-glycyrrhetinic acid, isogabrolide, liquirtic acid, 23- hydroxy glycyrrhetic acid, 24-hydroxy-11-deoyglycyrrhetic acid, glycyrrhetol, 21 a-hydroxy-isoglabrolide, 24-hydroxyliqultitic acid, liquiridiolic acid, 18a hydroxyglycyrrhetic acid, 28- hydroxy-glycyrrhetic acid, glabric acid and liquoric acid (Procter 1871; Alam, Hussain and Anwar 1982; Chauhan, Singh and Agarwal 1999; Deb and Mandal 1999, Cooke 1970; Housemen 1944-45; Hurst and Mckim 1983; Kinm *et al* 2000; Paik, Deb and Mandal 1999; Brieskron and Mahran 1960; Nam Ho *et al* 1981; Drozdz 1958; Canonica *et al* 1966; Canonica *et al* 1966; Canonica *et al.,* 1967; Canonica *et al* 1967; Canonica *et al* 1967; Canonica *et al* 1968, Canonica *et al* 1967; Elgamal *et al* 1975; Elgamal and Fayez, 1968; Elgamal *et al* 1965; George 1959; Hattori *et al* 1983; Hayashi *et al* 1998; Voss Klein and Sauer 1936), isoflavones; glyzaglabrin, glyzarin, formononetin, glabrene, glabrol and 3 hydroxyglabrol (Bhardwaj and Singh 1977; Bhardwaj *et al* 1977; Elgamal and Fayez 1968; Reiners 1966), coumarin; liqcoumarin (Bhardwaj *et al* 1976), flavanone; glabranine (Kattaev and Nikpnov, 1974), alkaloid; licorine (Humber *et al* 1954; Hyung, Tai amd Kang, 1969; Kang 1969), flavonoid glycoside; licurasidc, glyphoside and liquiritoside (Litvinenko and Kovalev 1966; Litvinenko 1966; Paris and Guillot 1955), flavanone glycoside; rhamnoliquiritin

(Van Hulle 1968) sterol; ß-sitosterol (Zayed Hassan and Elghamry 1964), phenolics; licopyrancoumarin, licoarylcoumarin, glicoricone and licofuranone, isoflavanes; hispaglabrin A, hispaglabrin B and glabridin, chalcone glycosides: licochalcone A, liquirtigenin and isoliquirtigenin (Gorecki 1991; Chen *et al* 1994), flavanones; licoflavanone, pincembrin, hydroxycoumarins; herniarin, glycycumarin, licopyranocoumarin and umbelliferone (Reiners 1964) and glycoflavonoids; saponaretin and vitexin (Litvinenko, and Kovalv 1967).

Pre-Clinical Studies: Anabolic (Malhotra and Ahuja, 1969), anti-anaphylactic (Gupta, Bhalla and Tangri 1970), antiarthritic (Gujral *et al* 1961; Tangri, *et al* 1969), anticancer (Agarwal, Wang and Mukhtar 1991; Eun, Kwon and Oh 1998; Shiota *et al* 2000). antidepressant (Dhingra and Sharma (2005), anti-leshmanial (Christensen *et al* 1994), antimalarial (Chen *et al* 1994), antioxidant (Belinky *et al* 1998), anti-nociceptive (Khaksa *et al* 1996), anti-platelet (Tawata *et al* 1992). antipyretic (Saxena and Bhalla 1968). antitussive (Anderson and Smith, 1961). anti-inflammatory (Tangri *et al* 1969; Inoue *et al* 1986; Jo, Kim and Ko 1986, Khaksa *et al* 1996). antimicrobial (Fukui Goto 1988; Fukai *et al* 2002), antiviral (Pompei *et al* 1980; Ito *et al* 1987; Hattori *et al* 1989; Aikawa Yoshiike, Ogawa, 1990; Badam 1997), choleretic (Raggi *et al* 1955), estrogenic (Muraviev and Kononikhina 1972), gastro protective (Saraswathi, 1986; Kharya 1995; Goso, Ogata, Ishihara and Hotta 1996). hepatoprotective (Nose *et al* 1994; Badam, 1997; Bean 2002). nootropic (Parle, Dhingra and Kulkarni 2004). thrombin inhibitor (Francishetti *et al* 1997), inhibitor of the antigranulomatous action of prednisolone (Kim, 1983). inhibitor of pituitary-adrenal stress response (Kraus 1958). mineralocorticoid activity (Kageyama, Suzuki and Saruta 1992). renoprotective (Sohan *et al* 2003) and stimulator of melanogenesis (Yokota *et al* 2000).

Clinical Studies

Antiviral

It has been demonstrated that glycyrrhizin has antiviral activity against DNA and RNA virus (VZV, HIV, Influenza A and B, types 1 and 2 herpes simplex and hepatitis B and C). *In vitro* and *in vivo*. Toxicity has not been found in experimental trials on different cell lines. Glycyrrhizin blocks the first steps on viral replication as well as the viron exit from the capside.

Hepatitis B

A multicenter, placebo-controlled study of 7 patients with late chronic viral hepatitis B showed that GL given 3 times a week for 12 months resulted in significant regression of the disease in 4 of the patients (Eisenburg 1992).

Hepatitis C

In a retrospective study, the long-term administration of glycyrrhizin in the treatment of chronic hepatitis C was effective in preventing hepatocellular carcinoma (Arase *et al* 1997).

Herpes simplex

In an open, controlled, clinical and multi-centric study, glycyrrhizin was evaluated for efficacy, safety and tolerability. 52 adult patients of both sexes (males 62.7%, females 37.3%) with age ranging from 18 to 65 years (mean 38.8) with active lesions of genital herpes simplex were included in this preliminary study. Patients the cases were enrolled under clinical, cytologic and immunologic criteria. A local application of the spray solution on the affected area was done twice a day for 5 days. Clinical efficacy was evaluated with no parametric methods. Total efficacy was observed in 95%, it was excellent in 65%, good in 25% and mild in 9%. The natural history was the same, but not so good were the control of pain and pruritus. The immuno-serological evolution was correlated with the clinical history. Local tolerability was of 94.1% and non-important adverse side events were observed (Vick Hidalgo, Zenon and Maritinez 2000).

Anti Hiv

A study from Japan explained the use of glycyrrhizin in 10 patients, who were given glycyrrhizin ranging from 150 to 225mg. At the end it was concluded that glycyrrhizin was more effective than placebo in preventing the occurrence of symptoms (Ito *et al* 1988).

Herpes Zoster

The authors studied the effect of glycyrrhizin on pain and HLA-DR antigen expression on CD8-positive cells in peripheral blood of herpes zoster patients in comparison with other antiviral agents (Aikawa, Yoshiike and Ogawa 1990).

Cytomegalovirus (possible hepatoprotective effect)

A study was done to evaluate effect of glycyrrhizin in children with liver dysfunction associated with cytomegalovirus infection. It was proposed that glycyrrhizin has inhibitory effect on immune mediated cytotoxicity against hepatocytes and on nuclear factor (NF)-kappa B, which activates genes encoding inflammatory cytokines in the liver (Numazaki *et al* 1994).

Skin diseases

Lichen Planus

A study reported effectiveness of glycyrrhizin for oral lichen planus in patients with chronic herpes simplex virus infection (Da Nagao *et al* 1996).

Melasma

A study reported effectiveness of topical liquiritin in the treatment of melasma (Amer and Metwalli 2000).

Obesity

A study reported effectiveness of Glycyrrhiza glabra in reducing body fat mass in healthy subjects (Armanini *et al* 2003).

Gastric ulcer

A clinical trial reported ulcer healing effect of deglycyrrhizinized liquorice in gastric ulcer (Turpie, Runcie and Thomson 1969).

Cardio protective

Plasma LDL from 10 normal subjects was found to be more resistant to oxidation following 2 weeks administration of 100 mg of licorice root per day (Fuhrman *et al* 1997).

Pharmacokinetics

Administration of *Glycyrrhiza glabra* after meals delays the time to peak concentration, but does not affect maximum concentration or area-under-the-curve. Elimination half-life is approximately 5 hours following intravenous administration. Primary route of excretion is in bile (Ichikawa *et al* 1986).

In an experimental study, a simple HPLC method was developed to study the pharmacokinetics of glycyrrhizin in the rat after bolus intravenous administration at a dose of 20, 50, or 100 mg/kg. The authors concluded that pharmacokinetics of glycyrrhizin is linear (Tsia, *et al* 1992).

Advers Drug Reactions: Sodium and water retention, and hypokalemia, hypertension (Walker, Edwards, 1994; Clyburn, DiPette, 1995), hyperprolactinemia (Werner, Brismar and Olsson 1979). hypersensitivity (Kuriyama *et al* 1975), may reduce human testosterone (Armanini *et al* 2004).

Teratogenicity: Teratogenic (Mantovani *et al* 1988).

BIBLIOGRAPHY

Adanin and Khalestkii, A.M. (1968). Synthesis of glycyrrhizic acid derivatives, I: Hydroxy derivatives. *Zh Obshch Khim* 37 (2): 329-331 *Chem Abstr* 68, 13202w.

Agarwal, R.; Wang, Z.Y. and Mukhtar, H. (1991). Inhibition of mouse skin tumor-initiating activity of DMBA by chronic oral feeding of glycyrrhizin in drinking water. *Nutr Cancer* 15 (3-4): 187-193.

Aikawa, Yoshiike, Ogawa (1990). Effect of glycyrrhizin pain and HLA-DR antigen expression on CD-8 positive cells in peripheral blood of herpes zoster patients in comparison with other antiviral agents. *Skin Pharmacol* 3: 268-271.

Alam, M.; Hussain, F. and Anwar, J. (1982). Crude drugs: Glycyrrhizin and its therapeutically active salts. *Pak J Sci* 34 (3-4): 61-64.

Anderson, J. and Smith, W.G., (1961) The antitussive activity of glycyrrhetinic acid and its derivatives. *J Pharm Pharmacol* 13: 396-404.

Arase, Y. *et al.* (1997). The long term efficacy of glycyrrhizin in chronic hepatitis C patients. *Cancer* 79: 1494-1500.

Armanini, D. *et al.* (2003). Effect of licorice on the reduction of body fat mass in healthy subjects. *J Endocrinol Invest* 26: 646-650.

Armanini, D., *et al.* (2004). Licorice reduces serum testosterone in healthy women. *Steroids* 69:763-766.

Badam, L. (1997). *In vitro* antiviral activity of indigenous glycyrrhizin, licorice and glycyrrhizic acid on Japanese encephalitis virus. *J Commun Dis* 29 (2): 91-99.

Bardhan, K.D. *et al.* (1979). Clinical trial of deglycyrrhizinised liquorice in gastric ulcer. *Gut* 19 (9): 779-782.

Bean, P. (2002). Silymarin and glycyrrhizin are best-known phyto-medicine for hepatitis C *Am Clin Lab* 21 (4): 19-21.

Belinky, P.A. *et al.* (1998). The antioxidative effects of the isoflavan glabridin on endogenous constituents of LDL during its oxidation, *Atheroscelrosis* 137(1): 49-61.

Bhardwaj, D.K. and Singh, R. (1977). Glyzaglabrin a new isoflavone from *Glycyrrhiza glabra. Curr Sci* 46 (21): 753.

Bhardwaj, D.K. *et al.* (1976). Liqcoumarin, A novel coumarin from *Glycyrrhiza glabra. Phytochemistry* 15 (7). 1182-1183.

Bhardwaj, D.K. *et al.* (1977). Glyzarin, new isoflavone from *Glycyrrhiza glabra. Phytochemistry*, 16 (3): 402-403.

Canonica, L. *et al.* (1966). Triterpenes of *Glycyrrhiza glabra*,11, Liquitric acid, *Gazz Chim Ital* 96 (6): 833-842.

Canonica, L. *et al.* (1966). Triterpenes of *Glycyrrhiza glabra*,111, Structure of isolabrolide. *Gazz Chim Ital* 96 (6): 843-851.

Canonica, L. *et al.* (1967). Triterpenes of *Glycyrrhiza glabra*, V11, 24-hydroxyliqultitic acid, liquiridiolic acid. *Gazz Chim Ital* 98: 712-728.

Canonica, L. *et al.* (1967). Triterpenes of *Glycyrrhiza glabra*, V, glycyr-rhetol, 21 α-hydroxy-isoglabrolide. *Gazz Chim Ital* 97 (10): 1347-1358.

Canonica, L. *et al.* (1967). Triterpenes of *Glycyrrhiza glabra*, V1, 23-hydroxy glycyrrhetic acid, 24-hydroxy-11 deoyglycyrrhetic acid. *Gazz Chim Ital* 97 (10): 1359-1369.

Canonica, L. *et al.* (1968). Triterpenes of *Glycyrrhiza glabra*, 1V, 18β hydroxyglycyrrhetic acid. *Gazz Chim Ital* 97 (5): 769-786.

Cermelli, C. *et al.* (1996). Activity of glycyrrhizin and its diastereoisomers against two new human herpes viruses: HHV-6 and HHV-7, *Phytotherapy Res* 10 (1): 527-528.

Chen, M, *et al.* (1994). Licochalcone A, a new anti-malarial agent, inhibits *in vitro* growth of the human malaria parasite *Plasmodium falciparum* and protects mice from *P. yoelii* infection. *Antimicrob Agents Chemother* 38:1470-1475.

Christensen, S.B., *et al.* (1994). An antileishmanial chalcone from Chinese licorice roots. *Planta Med* 60:121-123.

Clyburn, E.B. and DiPette, D.J. (1995). Hypertension induced by drugs and other substances. *Semin Nephrol* 15:72-86.

Cooke, M.K. (1970). Ammoniated glycyrrhizin: a useful natural sweetener and flavour potentiator. *Flavour Ind* 1 (12): 831-832.

Da Nagao Y., *et al.* (1996). Effectiveness of glycyrrhizin for oral lichen planus in patients with chronic HCV infection. *J Gastroenterol* 31: 691-695.

Deb, S. and Mandal, S.K. (1999). Standardization of Herbal Drugs Part 11, An estimation method of 18 ß-glycyrrhetinic acid obtained from glycyrrhizin in Glycyrrhiza glabra (*Yastimadhu*) by thin layer chromatography-densitometric method. *Indian Drugs* 36 (11): 687.

Dhingra, D. and Sharma, A. (2005). Evaluation of antidepressant-like activity of glycyrrhizin in mice. *Indian Journal of Pharmacology* 37 (6): 390-394.

Elgamal, M.H.A. and Fayez, M.B.E. (1972). Isolation of formononetin from the roots of *Glycyrrhiza glabra* L. collected locally. *Indian J Chem* 10 (1): 128.

Elgamal, M.H.A. *et al.,* (1965). Constituents of Local Plants-VI, Liquoric acid-A new triterpenoid from the roots of *Glycyrrhiza glabra* L. *Tetrahedron* 21 (8): 2019-2115.

Elgamal, M.H.A. and Fayez, M.B.E. (1968). Structure of glabric acid, a further triterpenoid constituent of *Glycyrrhiza glabra. Chem Abstr* 70: 29115x.

Eisenburg, J. (1992). Treatment of chronic hepatitis B. Part 2: Effect of glycyrrhizic acid on the course of illness. *Fortschr Med* 110(21): 395-398 28.

Eun, J.S.; Kwon, J. and Oh, C.H. (1998). Effect of glycyrrhizin on apoptosis of transplanted cells in mice. *Yakhak Hoeji*, 42(3) : 324-329.

Francishetti, I.M. *et al.* (1997). Identification of glycyrrhizin as thrombin inhibitor. *Biochem Biophys Res Commun* 235 (1): 259-263.

Fuggersberger-Heinz, R. and Franz, G. (1984). Formation of glycyrrhizinic acid in Glycyrrhiza glabra var. typica. *Planta Med* 50 (5): 409-413.

Fukui, H. and Goto, K. (1988). Two antimicrobial flavanones from the leaves of *Glycyrrhiza glabra*. *Chem Pharm Bull* 36 (10): 4174-4176.

Fukai, T., *et al.* (2002). Anti-Helicobacter pylori flavonoids from licorice extract. *Life Sci* 71: 1449-1463.

Fuhrman B., *et al.* (1997). Licorice extract and its major polyphenol glabridin protect low-density lipoprotein against lipid peroxidation: *In vitro* and *ex vivo* studies in humans and in atherotic apolipoprotein E-deficient mice. *Am J Clin Nutr* 66 (2): 267-275.

Goso, Y., Ogata, Y., Ishihara, K. and Hotta, K. (1996). Effects of traditional herbal medicine on gastric mucin against ethanol-induced gastric injury in rats. *Comp Biochem Physiol C Pharmacol Toxicol Endocrinol* 113: 17-21.

Groen, J. *et al.* (1952). Effect of glycyrhizinic acid on the electrolyte metabolism in Addison's disease, *J Clin Invest* 31 (1): 87-91.

Gujral, M.L. *et al.* (1961). Anti-arthritic activity of glycyrrhizin in adrenalectomised rats, *Indian J Med Sci* 15: 625.

Gujral, M.L. *et al.* (1961). Endocrinological studies on glycyrrhizin in rats, *Indian J. Med. Sci.* 15: 707.

Gupta, M.B., Bhalla. T.N. and Tangri, K.K. (1970). Anti-anaphylactic activity of glycyrrhetic acid, 6-bromquinazolone, imipramine and its congeners, *Jpn. J. Pharmacol* 20: 167.

Hattori, M. *et al.* (1983). Metabolism of glycyrrhizin by human intestinal flora, *Planta Medica* 48 (1): 38-42.

Hattori, T. *et al.* (1989). Preliminary evidence for inhibitory effects of glycyrrhizin on HIV replication in patients with AIDS. *Antiviral Res* 11: 255-262.

Houseman, P.A. (1944-45). Glycyrrhizin. *Sci and Cult* 10: 544.

Humber, L.G. *et al.* (1954). Licorice alkaloids, XXVIII. Constitution of lycorine and synthesis of its degradation products. *J Chem Soc* 4622-4630.

Hyung Tai, Kang (1969). Alkaloids of *Glycyrrhiza glabra*, *Chung-ang Vihak* 16 (6): 515-522.

Ichikawa T, *et al.* (1986). Biliary excretion and enterohepatic cycling of glycyrrhizin in rats. *J Pharm Sci* 75: 672-5.

Inoue, H. *et al.* (1986) Inhibitory effect of glyzyrrhetinic acid derivatives of lipo-oxygenase and prostaglandin synthetase. *Chem Pharm Bull* 34: 897.

Ito, M. *et al.* (1987) Inhibitory effect of glycyrrhizin on the *In vitro* infectivity and cytopathic activity of the human immunodeficiency virus (HIV (HLTY-III/LAV). *Antiviral Res* 7: 127-137.

Ito, M., *et al.* Mechanism of inhibitory effect of glycyrrhizin on replication of human immunodeficiency virus. *Antiviral Res* 10 (6): 289-98.

Jo, Y.S..; Kim, N.D. and Ko, K.H. (1986). The effects of licorice fraction and glycyrrhizin on prostaglandin synthetase activity of bull seminal vesicle, *Korean J Pharmacog* 17 (2): 107-112.

Kageyama, Y., Suzuki, H. and Saruta, T. (1992). Glycyrrhizin induces mineralocorticoid activity through alterations in cortisol metabolism in the human kidney. *Journal of Endocrinology* 135 (1): 147-152

Katsuya, I. *et al.* (1983). Prevention of the *In vitro* myelo-suppressive effect of lipopolysaccharide and hydrocortisone by glycyrrhizin. *Igaku no Ayumi* 125 (6): 554-556.

Kattaev, N.S. and Nikonov, G.K. (1972). Glabranine, a new flavanone from *Glycyrrhiza glabra Khim Pror Soedin* 8 (6): 805-806.

Khaksa, G. *et al.* (1996) Anti-inflammatory and anti-nociceptive activity of disodium glycyrrhetinic acid hemiphthalate. *Planta Med* 62 (4): 326-328.

Kharya, M.D. (1995). Tissue culture studies on a prospective antiulcer plant *Glycyrrhiza glabra. Indian J Nat Prod* 11 (1): 20-27.

Kim, C.J. (1983). The inhibitory effect of glycyrrhizin and bergenin on the antigranulomatous action of prednisolone, (32nd Annual Convention, *Pharm Soc* Korea, Seoul, Korea). *Yakhak Hoeji*, 27 (4): 374.

Kimura, M., *et al.* (1984). Blocking effects of blended paeoniflorin or its related compounds with glycyrrhizin on neuromuscular junctions in frog and mosue, *Jpn J Pharamacol* 36 (3). 275-282.

Kraus, S.D. (1958). glycyrrhizin-induced inhibition of the pituitary-adrenal stress response. *The Journal of Experimental Medicine* 108: 325-328,

Kso, Y. *et al.* (1984). Mechanism of antihepatotoxicvity activity of glycyrrhizin, I: Effect on free radical generation and lipid peroxidation, *Planta Med* 51 (4): 298-302.

Kuriyama, Y. *et al.* (1975). Hypersensitivity to glycyrrhizin a case report, *Med J Osaka Univ* 26 (1-2): 75-78.

Litvinenko, V.I. and Kovalv, I.P. (1966). Licuraside, a new flavonoid glycoside of *Glycyrrhiza glabra. Dokl Acad Nauk* SSSR, 169 (2): 347-350.

Litvinenko, V.I. and Kovalv, I.P. (1967). Glycoflavonoids of *Glycyrrhiza glabra* 1. Saponaretin and vitexine. *Khim Prir Soedin* 3 (1): 56-57.

Litvinenko, V.I. (1964). Chalcone glycosides of *Glycyrrhiza glabra, Dok Akad Nauk* 155 (3): 600-602.

Litvinenko, V.I. (1966). Glyphoside- new flavonoid glycoside obtained from *Glycyrrhiza glabra. Rastit Resur* 2 (4): 531-536.

Malhotra, S.C. and Ahuja, M.M.S. (1969). Efficiency of glycyrrhetinic acid as a steroidal agent, *Indian J Med Res* 57: 2232.

Mantovani, A. *et al.* (1988). Teratogenicity study of ammonium glycyrrhizinate in the Sprague-Dawley rat. *Food Chem Toxicol* 26 (5): 435-440.

Miyamura, M., *et al.* (1996). Properties of glycyrrhizin in Kampoo extracts including licorice root and changes in the blood concentration of glycyrrhetic acid after oral administration of Kampoo extracts. *Yakugaku Zasshi* 116: 209-16.

Moon, A. *et al.* (1995). Effect of *Glycyrrhiza glabra* roots and glycyrrhizin on the glucouronidation in rats, *Planta Med* 63 (2): 115-119.

Muraviev, A. and Kononikhina, N.F. (1972). Estrogenic properties of *Glycyrrhiza glabra* L. *Rast Resur* 8 (4): 490-497.

Nose, M. *et al.* (1994). A comparison of the antihepatotoxic activity between glycyrrhizin and glycerrhetinic acid. *Planta Med* 60 (2): 136.

Numazaki, K., *et al.* (1994). Effect of glycyrrhizin in children with liver dysfunction associated with cytomegalovirus infection. *Tohoku J Exp Med* 172 (2): 147-53.

Numuzaki, K., Umetsu, M. and Chiba, S. (1994). Effects of glycyrrhizin in children with liver dysfunction associated with cytomegalovirus infections. *Tohoku J Exp Med* 172: 147-153.

Paris, R.A. and Guillot (1955). Liquiritoside, flavonoside from root of licorice, *Glycyrrhiza glabra* L. *Ann Pharm Franc* 13 (9-10): 592-5.

Parle, M., Dhingra, D. and Kulkarni, S.K. (2004). Memory-strengthening activity of *Glycyrrhiza glabra* in exteroceptive and interoceptive behavioral models. *J Med Food* 7: 462-466.

Pompei, R., *et al.* (1980). Antiviral activity of glycyrrhizic acid. *Experientia* 36: 304.

Raggi, M.A. *et al.* (1955). The choleretic effects of licorice: identification and determination of the pharmacologically active components of *Glycyrrhiza glabra. Boll Chin Farm* 134 (11): 634-8.

Reiners, W. (1964). Coumarins and hydroxycinnamic acids from licorice roots. *Natur-wissenchaften* 51 (8): 193.

Saraswathi, P. (1986). Antiulcer activity of *Glycyrrhiza glabra* Int Conf on Trad Med Madras India Abstr No. D-47.

Saxena, R.C. and Bhalla, T. (1968). Antipyretic effect of glycyrrhetic acid and imipramine. *Jpn J Pharmacol* 18: 3563.

Shiota, G. *et al.* (2000). Inhibition of hepatocellular carcinoma by glycyrrhizin in diethylnitrosamine-treated mice. *Carcinogenesis* 20 (1): 59-63.

Sohan, E.J. *et al.* (2003). Protective effect of glycyrrhizin on gentamicin-induced acute renal failure in rats. *Basic and clinical pharmacology and toxicology* 93 (3): 116-7.

Suzuki, H. *et al.* (1983). Effect of glycyrrhizin on biochemical test in patients with chronic hepatitis. Double blind trial. *Asian Med J* 26: 423-438.

Tamura, Y. *et al.* (1979). Effects of glyzyrrhenitic acid and its derivatives on Delta5-reductase in rat liver. *Drug Res* 29: 647.

Tangri, K.K. *et al.* (1969). Biochemical study of anti-inflammatory and antiarthritic property of glycyrrhetic acid. *Biochem Pharmacol* 14:1277.

Tawata M. *et al.* (1992). Antiplatelet action of isoliquiritigenin, an aldose reductase inhibitor in licorice. *Eur J Pharmacol* 212 (1): 87-92.

Tsia, T.H. *et al.* (1992). Pharmacokinetics of glycyrrhizin after intravenous administration to rats. *J Pharm Sci* 81 (9): 961-3.

Turpie, A.G., Runcie, J. and Thomson, T.J. (1969). Clinical trial of deglycyrrhizinized liquorice in gastric ulcer. *Gut* 10: 299-302.

Van Hulle, C. (1968). Rhamnoliquitrin, a new flavanone glycoside from the roots of *Glycyrrhiza glabra* var *typica*. *Pharm Tijdschr* Belg 45 (7): 137-145.

Vick, F.R., Hidalgo, L.H., Zenon, M.C. and Maritinez, S. (2000). Local application of glycyrrhizin acid in genital herpes. *Rev Hosp M Gea Glz* 3 (4): 141-144.

Voss, W.; Klein, P. and Sauer, H. (1936). Glycyrrhizin, *Agnew Chem* 49: 556.

Walker, B.R. and Edwards, C.R. (1994). Licorice-induced hypertension and syndromes of apparent mineralocorticoid excess. *Endocrinol Metab Clin North Am* 23: 359-377.

Werner, S.; Brismar, K. and Olsson, S. (1979). Hyperprolactinaemia and licorice *Lancet* 1 (8111): 319.

Yamamura, Y. *et al.* (1992). Pharmacokinetic profile of glycyrrhizin in healthy volunteers by a new high-performance liquid chromato graphic method. *J Pharm Sci* 81 (10): 1042-1046.

Zhao, M.Q. *et al.* (1983). The preventive and therapeutic actions of glycyrrhizin, glycyrrhetic acid and crude alkaloids on experimental cirrhosis in rats, *Acta Pharm Sin* 18 (5): 325-331.

*Carbenoxolone is disodium salt of glycyrrhetinic acid hemisuccinate.

YELLOW MARSH MARIGOLD

Botanical Name: *Caltha palustris* Linn.

Natural Order: Ranunculaceae

Active Constituents: Alkaloids, triterpene saponins (Bhandari *et al* 1987), lactones: caltholid, palustrolid (Bhandari *et al* 1984) and ranunculus glucosides: ranunculin converting to proto-anemonine and anemonine (Bruni *et al* 1986).

Non-Clinical Studies: Anti-inflammatory (Anon: PDR for Herbal Medicines), hypolipidemic (Anon: PDR for Herbal Medicines).

Toxicity: Gastro-intestinal irritant.

BIBLIOGRAPHY

Anon (2000). PDR for Herbal Medicines. Medical Economics Company Inc, Montvale, NJ: 812-813.

Bhandari, P. *et al.* (1987). Triterpenoid saponins from *Caltha palustris. Planta Med* 53 (1): 98-100.

Bhandari, P. *et al.* (1984). Two non-triterpene lactones from *Caltha palustris. PH* 23 (8): 1699-1702.

Bruni, A. *et al.* (1986). Protoanemonine detection in *Caltha palustris. J Nat Prod* 49 (6): 1172-1173.

Part - II

Hypericum perforatum

1

Introduction and History

Synonyms

Amber touch-and-heal, Balm-of-warrior's wound, Devil's scorge, God's wonder plant, Herba hyperici, Klamath weed, Goat weed, Perforate St John's wort, Rosin rose, Saint-John's-wort, Touch and Heal and Witcher's herb (English), Blutkraut, Herrgottsblut, Hexenkraut, Johaniskraut and Walpurgiskraut (German), Herbe de milleprertuis (French), Hyufarikun (Arabia), Bassant (Hindi) and Khoontir (Kashmir).

Historical aspects

Hypercium perforatum Linn., popularly known as St. John's wort, has recently received attention of the medical world. The plant has been assigned names like herbal Prozac (trade name of selective serotonin reuptake inhibitor, fluoxetine), miracle antidepressant remedy and even some people have addressed it as quack medicine. The use of the plant for the treatment of mental illness is documented in history of herbal medicine.

Hypericum perforatum Linn., is member of family Guttiferae (Clausiaceae) or Hypercinaceae. According to Taskhtajan (1960), the morphological and chemical differences of the two families are insufficient to justify separating them.[1] The genus Hypericum is represented by 400 species and *Hypericum perforatum, Hypericum androsaemum, Hypericum japonicum, Hypericum angustifolium* and *Hypericum calycinum* are some notable plants.

Origin of the drug is controversial as far as Ayurveda is concerned. According to Nadkarni (1976) and Karnick (1994), *Hypericum perforatum* grows in temperate Western Himalayas and is known as 'Bassant.[2,3] Dymock, Warden and Hooper, in the book 'Pharmacographia Indica' have described that number of species of Hypericum are distributed in hilly parts of India but they have not mentioned the Ayurvedic usage of the plant.

Bombbardelli and Moraazzoni (1995), in their work reported Hypericum as wild weed growing along roadsides.[4] According to Upton (1997), Hippocrates was the first person to document the medicinal use of *Hypercium perforatum.*[5] Noted scholars like Galen and Dioscorides were aware about the medicinal use of *Hypercium perforatum.*[6,7] According to Reuter (1998), Paracelsus, the great physician, named *Hypercium perforatum* as 'arnica for the nerves'.[8] *Arnica montana* Linn. (Asteraceae) is widely used by Homeopathic doctors for curing the diseases of the nervous system. Gerard noted the vulnerary activity of *Hypercium perforatum* and according to him it is the best plant for healing chronic indolent ulcers. According to Silberman (1998), *Hypercium perforatum* was used by Italic people to drive away evil spirits.[9]

The Arabic Materia Medica describes a plant 'Hyufarikun' and according to Dymock, Warden and Hooper, it is equivalent to Saint John's wort of the Europe.

In Arabic language the meaning of the term 'Hyufarikun' is 'magic barley'. Euryphon, a Greek doctor (dated back to 288 BC) used the word Yperikon for the first time.[10] Linnaeus has thrown light on the meaning of world 'Hypericum'. It is derived from two words Hyper (top) and eikon (figure).[11] According to Heiligenstein and Guenmther (1998), the word Saint John's wort is given to plant because it was harvested on the occasion of St. John's day, which falls on 24th of the June.[6] In German language, the name for St. John's wort is hexenkraut meaning magic herb.[12]

Hooper, Warden and Dymock has documented the plant to be anthelmintic, diuretic and emmenagouge, when taken internally.

Externally it is used in the treatment of skin ulcers and rheumatism. Rasmussen (1998), reported about use of *Hypercium perforatum* as antidepressant and anxiolytic in European system of medicine[13]. Herbalists in America have used *Hypercium perforatum* in the treatment of nervous depression[14]. In folk medicine, *Hypercium perforatum* is used in the treatment of gastritis, sciatica, gout, rheumatism, malaria, edema and nocturnal enuresis.[15]

Today, *Hypericum perforatum* is one of the extensively studied medicinal plant. The active constituents of the plant have been identified and extracts have been standardized to marker compound. Clinical research has indicated usefulness of *Hypercium perforatum* in the treatment of mild to moderate depression and anxiety neurosis. Hypericin, a chemical present in the plant has shown antiviral activity and another constituent, hyperforin has been investigated as antibiotic. Although the constituent responsible for antidepressant activity of *Hypercium perforatum* is still to be identified, the plant has definite role to offer in depression.

Habitat

Hypercium perforatum is native to North Africa, West Asia and Europe. In United States, the herb grows abundantly in Northern California. The drug is imported from Eastern Europe.[5] In India, the plant grows in Western Himalayas and Shimla.[16]

Botany

Hypericum perforatum is a perennial, erect and glabrous herb. Roots are fusiform and branched. Stem is round having distinct lengthwise ridges. Leaves are oblong, opposite, sessile, blue-green, translucently dotted and having pale lower surface. Flowers are pentamerous and yellow in colour. Sepals are lanceolate and petals are black dotted. Stamens are in three bundles. The flowers on squeezing yield red coloured juice. Fruit is ovate and triangular capsule. Seeds are minute and cylindrical. The plant flowers-fruits between April and October.[5] According to Shakirova (1970), *Hypericum perforatum* should not be confused with rose of *Hypericum calycinum*, a common ornamental in America. Rose of *Hypericum calycinum* flowers and leaves are much larger than those of *Hypericum perforatum*.[17]

References

1. Taskhtajan, A.L. (1980). *Bot. Rev* 46: 225.

2. Nadkarni, K.M. (1976). *Indian Materia Medica*. Bombay: Popular Prakashan. 673.
3. Karnik, C.R. (1994). *Pharmacopoeial Standards of Herbal plants*. Vols.1-2. Delhi. Sri Satguru Publications. 1 : 189-92; 2 :125.
4. Bombardelli, E. and Morazzoni, P. (1995). *Hypericum perforatum*. *Fitoterpia*. 66 (1): 43-68.
5. Upton, R. (Ed.) (1997). *St. John's Wort Monograph*. American Herbal Pharmacopoeia.
6. Gunther, R.T. (1933). *The Greek Herbal of Dioscorides*. Hafner Pub. Co. (1968).
7. Alleyne, J. (1733). *A New English Dispensatory*. Tho. Astley, London.
8. Reuter, H.D. (1995). St. John's wort as an herbal antidepressant. *Eur J Herbal Med* Part 1:1(3):19-24; Part 11:1(4):15-21.
9. Silberman, H.C. (1996). Superstition and medical knowledge in an Italian herbal. *Pharmacy in History*. 38: 87-89.
10. Pickering, C. (1879). *Chronological History of Plants*, Little, Brown & Co., Boston.
11. Jaeger, E.C. (1972). A Source-Book of Biological Names and Terms. Charles C. Thomas, Springfield, IL.
12. Braun, R *et al.* (1997). *Standadardzulassungen fur Fertigaznemittel-Text and Komnentar*. Stuttgart: Deutscher Apotheker Verlag.
13. Rasmussen, P. (1998). St. John's wort-A review of its use in depression. *Australian J Med Herbalism* 10(1): 8-13.
14. Leung, A.Y. and S. Foster (1996). *Encyclopedia of Common Natural Ingredients Used in Food, Drugs and Cosmetics*, 2nd ed. New York: John Wiley and Sons.
15. Hauselhuber, H. Kleinachimdt and S. Knust von Wedel (1969). *Hippoktares*. 40, 105.
16. Gaind K.N. and Ganjoo, T.N. (1959). Antibacterial principle of *Hypericum perforatum* L. *Indian Journal of Pharmacy*. 21:172-175.
17. Shakirova, K.K., *et al.* (1970). Antimicrobial properties of some species of St. John's wort cultivated in Uzbekistan *Mikrobiol. Zh. (Kiev.)* 32: 494-497.

2

Chemical Composition

Hypercium perforatum is a true example of a chemical factory. Medicinal plants are small chemical factories containing number of constituents. Nahrstedt and Butterweck (1998) divided chemical constituents of *Hypercium perforatum* into six distinct classes: naphtodianthrones, flavonoids, phloroglucinols, xanthones, phenolic acids and terpenes. According to Erdelmeier (1998), *Hypercium perforatum* contains naphtodianthrones derivatives, flavonoids, phloroglucinols, xanthones and essential oil.[1] Heiligenstein and Guenther (1998) proposed that at least ten constituents are responsible for pharmacological activity of *Hypercium perforatum*.

1. Napthodianthrones

Napthodianthrones are derivatives of anthracene. The napthodianthrones of *Hypercium perforatum* includes hypericin, pseudohypericin and isohypericin. The fresh plant contains protohypericin and protopseudohypericin and in presence of light, they are converted into hypericin and pseudohypericin.[2] Hypericin ($C_{30}H_{16}O_8$) is soluble in acetone and pyridine.

Newallet *et al.*, (1996), writers that the total amount of napthodianthrones ranges from 0.05-0.15%.[3] According to Upton (1997) the hypericin content of dried flowers can be upto1.8%. *Hypericum perforatum* extract used in most clinical trials has been standardized to contain at least 0.3% hypericin. However hypericin is now a days regarded as marker compound in *Hypericum perforatum* extracts.[5]

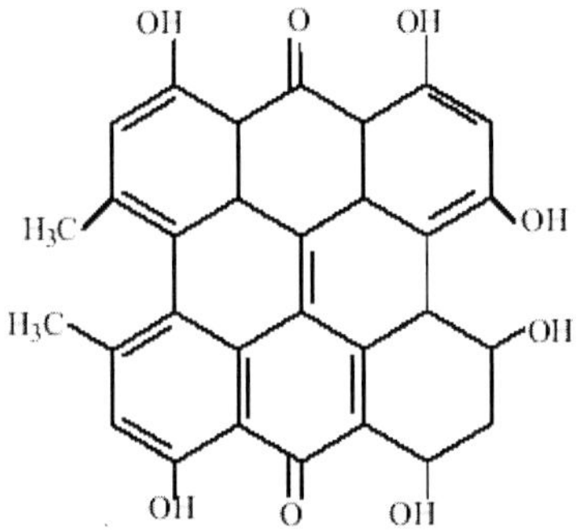

Fig. 1. Structure of Hypericin

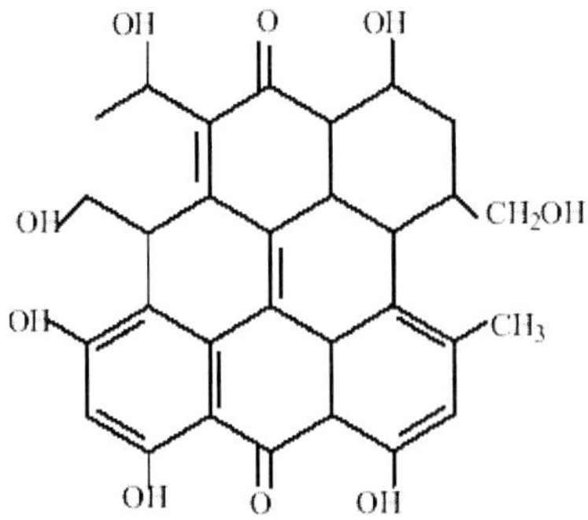

Fig. 2. Structure of Pseudohypericin

2. Acyphloroglucinols

According to Erdelmeier (1998), acylphloroglucinols are found in fresh herb of *Hypericum perforatum.*[1] Main acylphloroglucinol is hyperforin (C_{35} H_{52} O_4). Hyperforin is soluble in DMSO and organic solvents.

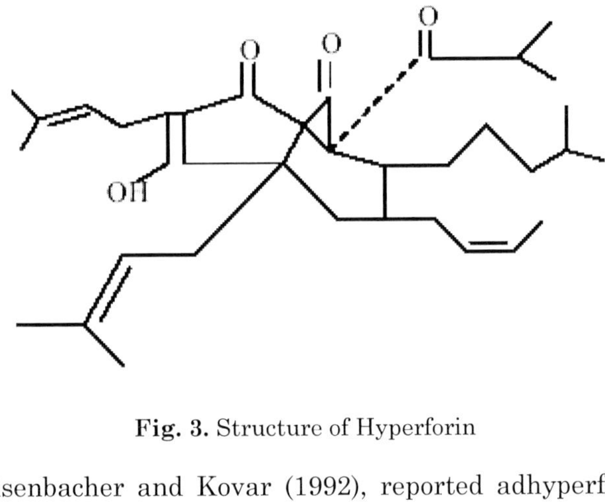

Fig. 3. Structure of Hyperforin

Maisenbacher and Kovar (1992), reported adhyperforin (see figure 4) from *Hypericum perforatum.*[6]

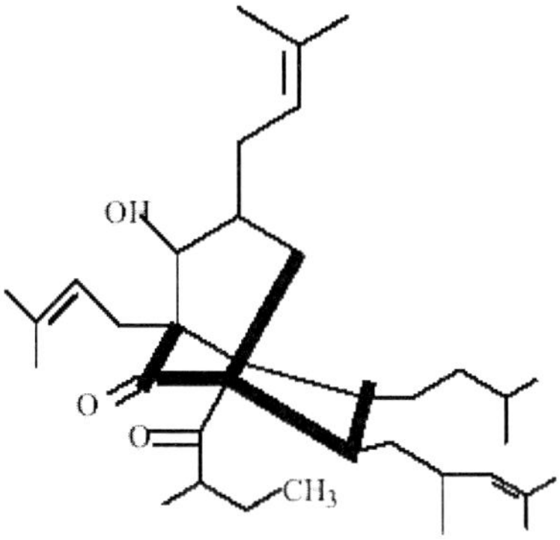

Fig. 4. Structure of Adhyperforin

According to Wagner and Bladt (1994), wild crafted *Hypericum perforatum* from Germany, dried at 30°C, was found to contain 2.8% hyperforin.[5] Rucker *et al.* (1992) reported a new compound containing a hyperforin and a sesquiterpene moiety from leaves and stems of *Hypericum perforatum*.[6] Erdelmeier (1998) states that hyperforin is highly lipophilic.[1] When exposed to air and light, hyperforin degrades rapidly. Chatterjee *et al.* (1998) explained that if *Hypericum perforatum* is extracted with supercritical carbon dioxide, higher amount of hyperforin can be obtained.[7]

Vugdelija *et al.* (2000) isolated phloroglucinol from aerial parts of *Hypericum perfortatum*.[8]

3. Flavonoids

Hyperoside (hyperin), quercetin, isoquercetin, rutin, methy-hespericin, iso-quercitrin, quercitrin, I-3/II-8-biapigenin (amento-flavone), kaempferol.[9,10] Butterweck *et al.* (2000), reported miquelianin and astilbin from *Hypericum perforatum*.

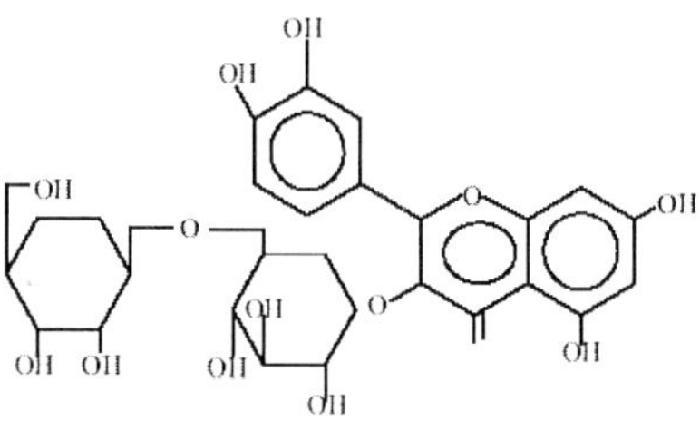

Fig. 5. Structure of Rutin

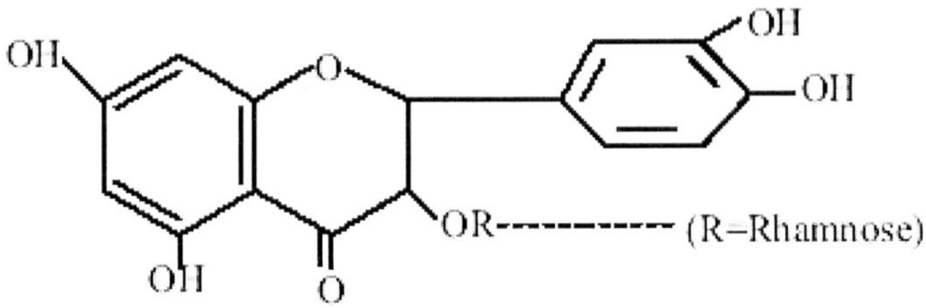

Fig. 6. Structure of Quercitrin

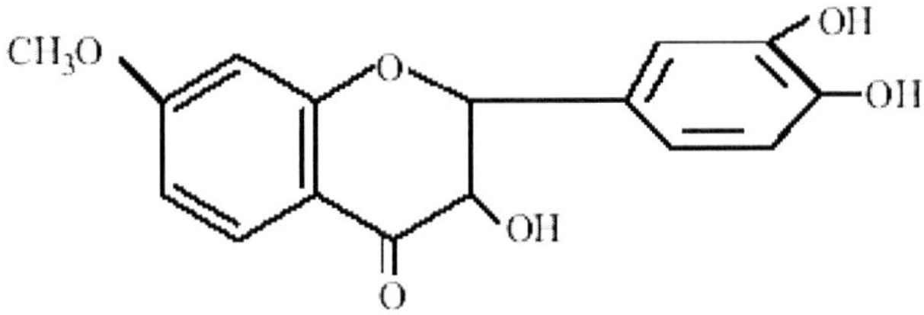

Fig. 7. Structure of Rhamnetin

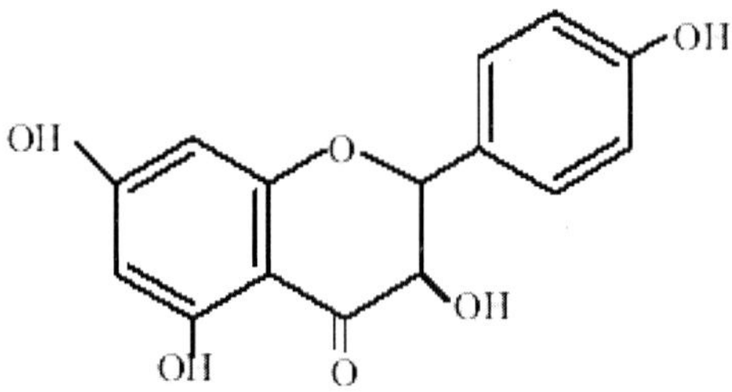

Fig. 8. Structure of Kaempferol

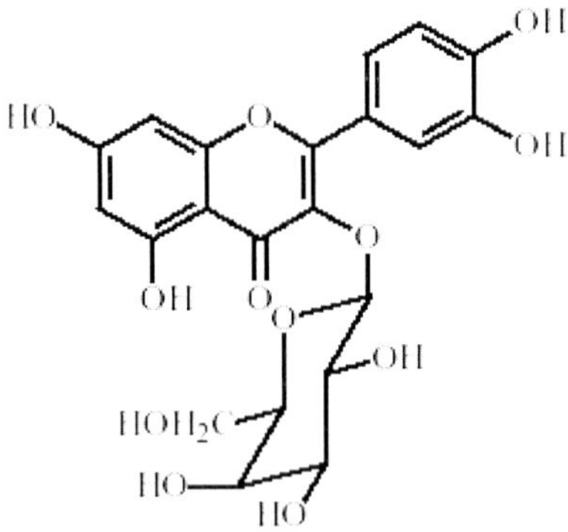

Fig. 9. Structure of Hyperoside

Baureithel *et al.* (1997) reported pharmacological activity of 1-3/II-8-biapigenin (amentoflavone) from *Hypericum perforatum*.[11]

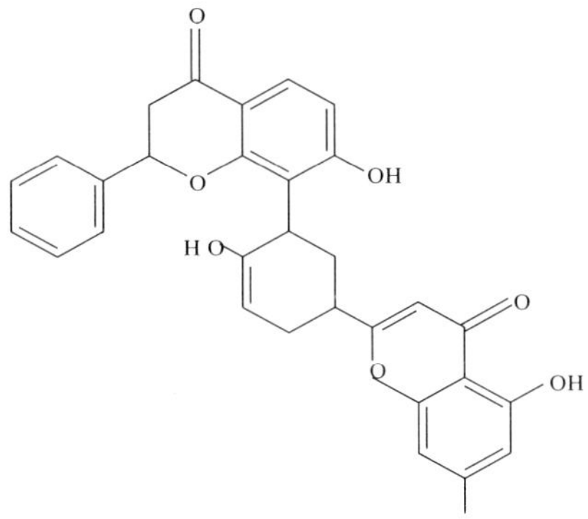

Fig. 10. Structure of Amentoflavone

4. Flavanols

Catechin type tannins and condensed proanthocyandins (catechin, leucocyanidin and (-)-epicatechin.[12]

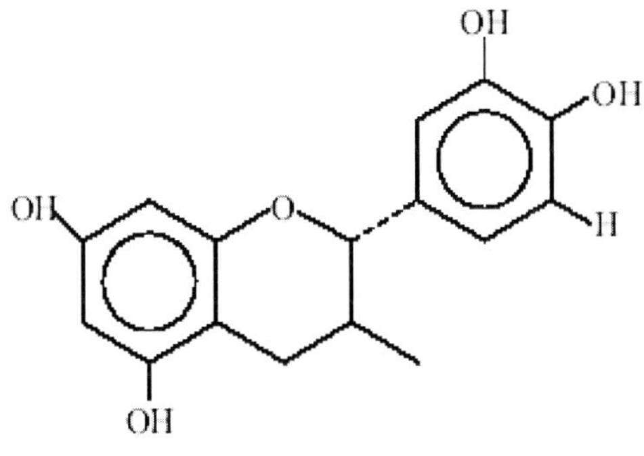

Fig. 11. Structure of Catechin

5. Xanthones[13]

6. Sterols[14]

β-sitosterol.

7. Vitamins [4]

Vitamin A and C.

8. Terpenoids

Monoterpenes (α-pinene, β-pinene, myrecene, limonene) and sesquiterpenes (caryophyllene and humulene).[15]

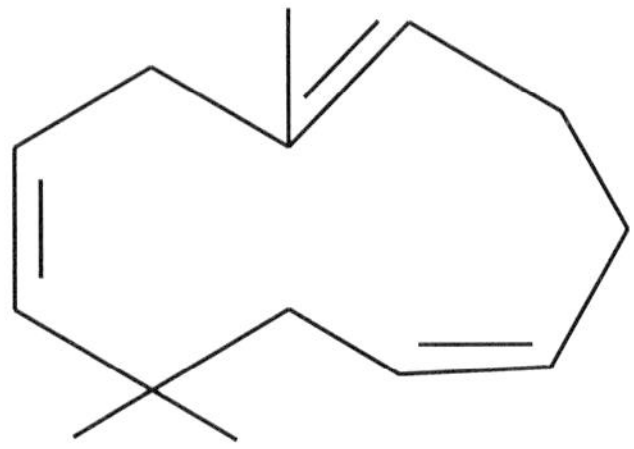

Fig. 12. Structure of Humulene

9. Carotenoids [16]

10. n-Alkanes[17]

11. n-Alkanols [18]

12. Tannins [19]

13. C50-hyrdoxyperoxide (hydroperoxycariforin) [20]

Telekova *et al.* (2000) studied the quantitative changes of active compounds in the flowers of *Hypericum perforatum*. The plant was harvested and flower development phase with respect to active compounds was examined. With help of HPLC, secondary metabolites during various stages of flower development were studied. The content of napthodianthrones, acylphloroglucinols and flavonoids increased from bud phase (0.29%, 2.47% and 0.80% respectively) to open position of flower (1.04%, 6.60% and 4.23% respectively). The content of napthodianthrones and flavonoids decreased in unripe ruits (0.11% and 0.08% respectively). Increase in amount of hyperforin (8.07%) was observed in the fruits. Amentoflavone content increased in the small buds but decreased in the fruits. Rutin was not observed in the samples.[21]

Adam *et al.* (2002) investigated the biosynthesis of hyperforin in *Hypericum perforatum*. The process involves five isoprenoid units, which originate via deoxyxylulose pathway. The phloroglucinol originates via polyketide pathway.[22] According to Verotta (University of Italy), acylphloroglucinols have a common 1, 3, 5-

trihydroxy-benzene core of polyketide origin. He considers acyl-phoroglucinols as lead compounds for pharmacotherapy of degenerative disorders.[23]

Walker (1999) has given beautiful explanation of hypericin biosynthesis.[24] Piovan *et al.* (2004) investigated sepals of *Hypericum elodes* with electro spray ionization mass spectrometry (ESI-MS/MS). The sepals contain red glands (stipitate glands with red-coloured heads). Piovan states that occurrence of hypericin in the red glands of *Hypericum elodes* is solid evidence for the taxonomic position of the section Elodes within the genus Hypericum. He further says that the ability of carrying out the biosynthetic pathway leading to the naphthodianthrone compounds, rather than the absolute amounts produced, should be regarded as a chemical marker of the phylogenetically more advanced sections of genus Hypericum. The phloroglucinol derivatives (hyperforin and adhyperforin, which so far found only in *Hypericum perforatum)*, were also detected and evidence for their localization in the sepal secretory canals with large lumen.[25]

Tolonen *et al.* (2002) described a high performance liquid chromatography method with photodiode array for prompt determination of napthodianthrones and acylphloroglucinols in extracts of *Hypericum perforatum.*[26] Wolfender *et al.* (2003), described structures of isomeric oxidized form of hyperforin with application of HPLC-NMR and HPLC-MS.[27] Balogh described advantage of multispectal techniques for detection of hypericin.[28]

References

1. Erdelmeier, C.A.J (1998). Hyperforin, possibly the major non-nitrogenous secondary metabolite of *Hypericum perforatum* L. *Pharmacopsychiatry.* 31: 2-6.

2. Bruneton, J. (1995). *Pharmacognosy, Phytochemsitry, Medicinal Plants.* Paris. Lavoisier Publishing.

3. Newall, C.A., Anderson, L.A. and Phillipson J.D. (1996). *Herbal Medicines: A Guide for Health Care Professionals.* London: The Pharmaceutical Press.

4. Upton R (Ed.). (1997). *St. John's Wort Monograph.* American Herbal Pharmacopoeia.

5. Wagner, H. and Bladt, S. (1994) Pharmaceutical quality of hypericum extracts. *Journal of Geriatrics, Psychiatry and Neurology* 7: 65-68.

6. Maisenbacher, P. and Kovar K.A. (1992). Adhyperforin - a homologue of hyperforin from *Hypericum perforatum*. *Planta Medica* 58: 291-293.

7. Chatterjee, S.S., Noldner, M., Koch, E. and Erdelmeier, C. (1998). Antidepressant activity of *Hypericum perforatum* and hyperforin - the neglected possibility. *Pharmacopsychiatry* 31: 7-15.

8. Suzana Vugdelija, Vlatka Vajs, Snezana Trifunovic, Dejan, Djokovic and Slobodan Milosavljevic (2000). A New Heterocyclization Product of Adhyperforin form *Hypericum perforatum* (St. John's wort). *MPDI* 5, M158.

9. Berghoefer, R. and Hoelzl J. (1987). "Biflavonoids in *Hypericum perforatum*. Part 1. Isolation of 13, II8- biapigenin," *Planta Med* 53: 216-17.

10. Koget, T.A. (1972). "Determination of the amount of quercitrin in *Hypericum perforatum*," *Khim. Prir. Soedin* (2):242-3 (CA77: 45514b).

11. Baureithel, K.H., Berger Buter, K., Engesser, A., Bukard, W. and Schaffner, W. (1999). Inhibition of benzodiazepine binding *in vitro* by amentoflavone, a constituent of various species of Hypericum. *Pharmaceutica Act Helvetiae* 72: 153-7.

12. Kitanov, G. 1983. "Determination of the absolute configuration of catechins isolated from *Hypericum perforatum*," *Farmatsiya* (Sofia) 33: 19-22.

13. Nielsen, M. and P. Arends. (1978). *Phytochemistry* 17: 2040.

14. Mathis C. and Ourisson G. (1964). "Etude Chimio-taxonomique du genre Hypericum-V.", *Phytochemistry* 3: 379.

15. Sticher, O. (1977). "Plant mono-, di- and sesquiterpenoids with pharmacological or therapeutical activity," in *New Natural Products and Plant Drugs with Pharmacological, Biological or Therapeutical Activity*, ed. by H. Wagner & P. Wolff, Springer-Verlag, NY.

16. Costes, C. and Chantal, T. (1967). Carotenoid pigments of the petals of the inflorescence of St. John's wort (Hypericum perforatum). *Ann Physiol Veg* 9: 157-77.

17. Brondz, I., *et al.* (1983). n-Alkanes of *Hypericum perforatum*: a revision, *Phytochemistry* 22: 295-6.

18. Brondz, I. and Greibrokk T. [d.m.]. "n-1-alkanols of *Hypericum perforatum*". *Journal of Natural Products* 46: 940

19. Duke, J.A. (1985). *Handbook of Medicinal Herbs*. Boca Raton, FL: CRC Press, pp. 242-243.

20. Rucker G., Manns, D., Hartmann, R. *et al.* (1995). A C50-hyrdoxy-peroxide from *Hypericum perforatum*. *Archive der Pharmazie* 328: 725-730.

21. Tekelova D. *et al.* (2000). Quantitative Changes of Dianthrones, Hyperforin and Flavonoids Content in the Flower Ontogenesis of *Hypercium perforatum*. Thieme: *Planta Medica*-Letters 8

22. Adam, P., Arigoni, D., Bacher, A. and Eisnreich, W. (2000). Biosynthesis of hyperforin in *Hypericum perforatum*. *Med Chem* 10; 45 (21): 4786-4793.

23. Verotta, L. (2001). Are phlorogucinols as lead structures for the treatment of degenerative diseases? Dipartimento di Chimica Organica e Industriale, Universita degli Studi di Milano, via Venezian 21, 20133 Milano, Italy.

24. Loren, W. and Walker, A. (1999). Review of the Hypothetical Biogenesis and egulation of Hypericin synthesis via the Polyketide Pathway in *Hypericum perforatum* and Experimental Methods Proposed to Evaluate the Hypothesis. Portland State University.

25. Piovan, A., Filippini, R., Caniato, R., Borsarini, A., Bini Maleci, L. and Cappelletti, E.M (2004). Detection of hypericins in the "red glands" of *Hypericum elodes* by ESI-MS/MS. *Phytochemistry* 65(4): 411-414.

26. Tolonen, A., Hohtola, A. and Jalonen, J. (2002). Fast high-performance analysis of napthodanthrones and phloroglucinols from *Hypericum perforatum* samples. *Phytochemical Analysis* 14(5): 306-309.

27. Wolfnder L., Verotta L., Belvisi N., Fuzzati and Hostettmann, K, (2003). Structural investigations of isomeric oxidises forms of hyperforin by HPLC-NMR and HPLC-MS. *Phytochemical Analysis* 14(8): 290-297.

28. Balogh, M.P. and Jeanne, B. Li. (2000). HPLC analysis of hypericin with PDA and MS detection, LC-GC, 17(6): 556-562.

3

Hypericum perforatum and Cardiovascular System

Negrash *et al.* (1969), in his experiments on the isolated frog heart, found that imanine and water soluble imanine caused systolic arrest at concentrations as low as 1 times 10 (-5).[1] Imanine and water-soluble imanine, from Hypericum extracts were reported to lower blood pressure after intravenous use in rabbits.[1]

Melzer *et al.* (1988), reported contractions of the isolated guinea-pig heart or isolated porcine coronary arteries induced by histamine and potassium chloride were inhibited by the proanthocyanidin containing components of *Hypericum perforatum*. The authors concluded that an inhibitory action of proanthocyanidin containing components on cellular phosphodiesterase might be responsible for the mechanism of action.[2]

Melzer *et al.* (1991), reported cardio protective activity of a procyanidin fraction of *Hypericum perforatum*. The procyanidin resulted in increased arterial flow in animals. Further the procyanidin demonstrated antagonistic effect on histamine induced arterial contractions.[3]

References

1. Negrash AK *et al.* (1969) Comparative study of chemotherapeutic and pharmacological properties of antimicrobial preparation from common St. John's wort. *Fitontsidy Mater Soveshch* 198-200.

2. Melzer R. *et al.* (1988) Proanthrocyanidins from *Hypericum perforatum* - effects on isolated pig coronary arteries. *Plant Med* 54: 572-573.

3. Melzer, R, Fricke, U. and Holzl, J. (1991). Vasoactive properties of procyanidins from *Hypercium perforatum* L. in isolated porcine coronary arteries. *Arzneimittel-Forschung/ Drug Research* 41:481-483.

4

Hypericum perforatum and Integument System

Hypericum perforatum has definite place in traditional medicine, as far treatment of skin diseases is concerned. Wound-healing activity of *Hypericum perforatum* has been highlighted by ancient healers.

Negrash *et al.* (1969) reported that a water-soluble imanine from *Hypericum perforatum* demonstrated antibacterial activity. It was superior to sulphanilamide.[1]

Brondz *et al.* (1982) reported antibacterial activity of hyperforin against *Staphlococcus aureus.*[2]

Khosa & Bhatia (1982) extracted hot water extract and a steam distillation derived volatile oil *Hypericum perforatum* L. The authors reported antifungal activity of both preparations. The extracts were tested against *Microsporum gypseum, Trichophyton rubrum, Aspergillus flavus, Curvularia lunato* and *Fusarium vasiinfectum*. The inhibition index of the hot water extract was 10 milligrams/ milliliter (mg/mL) for the *Microsporum gypseum* and *Trichophyton rubrum* species and 6 mg/mL for the others. The oil showed inhibition in all species tested with a 4% solution in 25% alcohol.[3]

Kolesnikova (1986) states that extracts prepared from *Hypericum perforatum* have more antibacterial activity as compared to decoctions.[4]

Sakar *et al.* (1988) reported antibacterial activity of Novoimanine, found in *Hypericum perforatum* as an effective topical agent against *Staphlococcus aureus.*

Schempp *et al.* (1990) investigated the immunomodulatory activity of topical preparations of *Hypericum perforatum.* The researchers investigated the alloantigen presenting function of human epidermal cells exposed to ointment of *Hypericum perforatum in vivo* in a mixed epidermal cells lymphocyte reaction. The effect of *Hypericum perforatum* was compared with immunosuppressive effect of solar-stimulated radiation. Hyperforin was purified and its possible role in the immunomodulatory activity of topical preparations of *Hypericum perforatum* was discussed. The researchers then investigated the effect of hyperforin on the proliferation of peripheral blood mononuclear cells *in vitro.*

When the results were compared, it was concluded that topical preparations of *Hypericum perforatum* caused significant suppression of mixed epidermal cells lymphocyte reaction as compared to the untreated group. The combination of ointment of *Hypericum perforatum and* solar-stimulated radiation was not significantly different from either treatment with ointment of *Hypericum perforatum.* The researchers concluded that topical preparations of *Hypericum perforatum* and hyperforin both have potential for treating inflammatory conditions of the skin.[6]

In a comparative study, Rao *et al.* (1991) evaluated wound healing effects of *Hypericum perforatum* and *Calendula officinalis* (marigold). Tinctures of fresh leaves and flowering tops of both the plants were used for topical application. The authors reported epithelization within fifteen days in *Hypericum* perforatum group and 16.5 days in *Calendula officinalis* group. The breaking strength of wounds was greater in *Hypericum* perforatum and *Calendula officinalis* group as compared to central group. However, in *Hypericum* perforatum group significant difference in granuloma weight was not observed.[7]

Smyshliaeva *et al.* (1992) reported modification of a radiation lesion in animals with an aqueous extract of *Hypericum perforatum.* The plant has been shown to decrease the intensity level of enzymatic and non-enzymatic processes of lipid peroxidation of rat liver microsomes *in vitro* and *in vivo.* In the mucus of this intestine *in vivo* the increase of the number of non-enzymatic SH-groups was observed.[8]

References

1. Negrash, A.K. *et al.* (1969). Comparative study of chemotherapeutic and pharmacological properties of antimicrobial preparation from common St. John's wort. *Fitontsidy Mater Soveshch.* 198-200.
2. Brondz, I. *et al.* (1983). n-Alkanes of *Hypericum perforatum*: A revision. *Phytochemistry* 22: 295-296.
3. Khosa R.L. and Bhatia, N. (1982). Antifungal effects of *Hypericum perforatum* Linn. *J Sci Res Plant & Med. 3:49*-50.
4. Kolesnikova, A.G. (1986). Bacterial and immunocorrective properties of plant extracts. *Zh Mikrobiol Epidemiol Immunobiol* 3:75-78.
5. Sakar, M.K. *et al.,* (1988). Antimicrobial activities of some Hypericum species growing in Turkey. *Fitoterapia* 59: 49-52.
6. Schempp, C.M., Winghofer, B., Ludtke, R., Simon-Harrhaus B., Schopf, E. and Simon, J.C. (2000). Topical application of St John's wort (*Hypericum perforatum* L.) and its metabolite hyperforin inhibit the allostimulatory capacity of epidermal cells. *Br J Dermatol.* 142 (5): 979-84.
7. Rao, S.G. *et al.* (1991). *Calendula* and *Hypericum*: two homeopathic drugs promoting wound healing in rats. *Fitoterpia* 6: 508-510.
8. Smyshliaeva, A.V., Nguyen, L.H., Kudriashov and Iu, B. (1992). The modification of a radiation lesion in animals with an aqueous extract of *Hypericum perforatum* L. 2. *In: Biol Nauki.* (4):9-13.

5

Hypericum perforatum and nervous system

In recent year, we have witnessed increasing popularity of *Hypericum perforatum* for the treatment of mild to moderate depression. Several articles and papers have been published favoring the use of *Hypericum perforatum* as antidepressant medication. Some studies have quoted the efficacy of *Hypericum perforatum* equivalent to standard antidepressants. More so, the herb has been used for treating depression and anxiety neurosis since ancient times. Market analysis has also shown that in Germany alone, in 1999, *Hypericum perforatum* was prescribed twenty times more than fluoxetine, the selective serotonin reuptake inhibitor. The physicians also reported favorable antidepressant profile of *Hypericum perforatum*. Following description will narrate work done on antidepressant action of *Hypcricum perforatum* in the last twenty years.

Early studies

1. Muldner and Zoller (1984) investigated the antidepressant action of *Hypercium perforatum* in six women in ago group of 55-65 years. They were treated with a commercial preparation of *Hypercium perforatum*. The effect of the drug on the effect on the excretion of urinary metabolites of nor adrenaline and dopamine was measured. The researchers observed a significant increase in 3-methoxy-4-hydroxyphenylglucol, which acts as a

signal for onset of antidepressant action of the drug. Nine more patients were treated with *Hypericum perforatum* and the clinical influence on the depression was measured during a period of 4-6 weeks, using definite assessment tools SCAG (Clinical assessment geriatric scale) and DSI (Depression Status Inventory). There was definite improvement in symptoms of depression including anxiety, loss of interest, hypersomnia (excess of sleep), anorexia (loss of appetite), depression regularly worse in the morning, insomnia (lack of sleep) and feeling of worthlessness.[1]

2. Okpanyi and Weischer (1987) studied the antidepressant action of extract of *Hypericum perforatum* with known concentrations of hypericin, in several models, generally accepted as screening methods in experimental animal studies for the recognition of antidepressant activity. Hypericum extract enhanced the exploratory activity of mice in a foreign environment. Hypericum extract prolonged the narcotic sleeping time dose-dependently. Hypericum extract within a narrow dose range exhibited reserpine antagonism. Hypericum extract enhanced significantly the activity of mice in the water wheel test. Prolonged daily administration of Hypericum extract decreased aggressiveness in socially isolated male mice.[2]

3. Schmidt and Sommer (1993) studied the efficacy of *Hypercium perforatum* (LI 160), in a placebo-controlled, randomized, double-blind trial. The study included patients suffering from mild to moderate depression. The group receiving Hypericum extract showed 66.6% improvement in symptoms of depression as compared to 26.7% with placebo. No adverse reaction was reported.[3]

4. Sommer and Harrer (1994) in a placebo-controlled double-blind study examined the effectiveness of a preparation of *Hypercium perforatum* (3 x 300 mg) or placebo in 105 mildly depressed patients. The therapy phase lasted for four weeks. Again the effectiveness of the treatment was assessed according to the Hamilton Depression Scale after two and four weeks. The group receiving Hypericum extract showed 67% improvement in symptoms of depression as compared to 28% with placebo.[4]

5. Muller and Rossol (1994) investigated the influence of extract of *Hypercium perforatum* on the expression of serotonin receptors using a neuroblastoma cell line to establish a model for the regulation of neurotransmitters by immunologically active compounds such as cytokines. The cells were incubated with extract of *Hypercium perforatum* in kinetic form for 2, 4, 6, 8, and 10 hours, and then washed. The serotonin receptor expression analysis was compared to that of a placebo control solution. The neuroblastoma cells showed a clearly reduced expression of the serotonin receptors under treatment with hypericum extract. First stimulation experiments with interleukin-1 (IL-1) and hypericum extract suggest that a further reduction of the serotonin receptors is possible when IL-1 is added.[5]
6. Thiele, Brink and Ploch (1994) investigated the effect of extract of *Hypercium perforatum* on the stimulated cytokine expression *in vitro* in a whole blood culture system. Blood samples were taken from five healthy volunteers and four depressive patients. The release of interleukin-6 (IL-6), interleukin- 1 beta (IL-1 beta) and tumor necrosis factor-alpha (TNF-alpha) was measured quantitatively after an incubation time of 24 hours on microtiter plates. A massive suppression of the interleukin-6 release was found for PHA-stimulated hypericum extract.[6]

Monoamine oxidase inhibiting activity of Hypericum perforatum

1. Bladt and Wagner (1994) studied the monoamine oxidase inhibiting activity of six fractions and three pure constituents of *Hypericum perforatum in vitro* and *ex vivo*. The objective of the study was to understand the antidepressant mechanism of *Hypericum perforatum*. Rat brain homogenates were used as the *in vitro* model, while the *ex vivo* analysis was performed after intraperitoneal application of the test substances to albino rats. The extract and all the fractions resulted in significant inhibition of MAO-A. The fraction rich in flavonoides showed an inhibition of 39% and all other fractions

demonstrated less than 25% inhibition. Hypericin failed to produce significant inhibition of MAO-A. The authors concluded that the antidepressant action of *Hypericum perforatum* can not be explained in terms of monoamine oxidase inhibiting activity. Also the antidepressant activity can not be assigned to single constituent as six fractions resulting in inhibition of monoamine oxidase.[7]

2. Thiede and Walper (1994) investigated the influence of total extract of *Hypericum perforatum*, hypericin, and hypericum fractions on the monoamine oxidase (MAO) and catechol-O-methyl transferase (COMT) in several concentration steps. An inhibition of monoamine oxidase was observed with following concentrations (extract correlated to a mean molecular value of 500): hypericin to 10(-3) mol/L, hypericum total extract to 10(-4) mol/L, one extract fraction up to 10(-5). A COMT inhibition could not be shown for hypericin, with hypericum extract to 10(-4) mol/L and with two extract fractions also up to 10(-4) mol/L. The monoamine oxidase inhibiting fraction contained napthodianthone (hypericins) and flavonols. The catechol-*O*-methyl transferase-inhibition fraction was mainly constituted of flavonols and xanthones. However the researchers concluded that the concentrations of inhibition shown were not sufficient to explain the clinically proven antidepressive effect of *Hypericum perforatum* with regard to the inhibition of MAO activity.[8]

Hypericum extract vs maprotiline

1. Johnson, Ksciuk, Woelk, Sauerwein-Giese and Frauendorf, (1994) in a randomized double-blind study, compared the effect of extract of *Hypericum perforatum* with maprotiline in 24 healthy volunteers. The investigations included measurements of resting electroencephalography (EEG) as well as visual and acoustic evoked potentials. In resting EEGs, both *Hypericum perforatum* and maprotiline revealed oppositely directed changes in the theta frequencies, and mainly similarly directed changes in alpha and beta frequencies. Measurements of evoked potentials in the theta and beta frequencies supported these results. The observations indicated

improved cognitive functions with the treatment of extract of *Hypericum perforatum*.[9]

2. Harrer, Hubner and Podzuweit, (1994) in a multi-center double-blind study, investigated the effect and tolerance of the extract of *Hypericum perforatum* (LI 160) in comparison to maprotiline. The study included 102 patients with depression. The patients received, over a period of 4 weeks, either 3 x 300 mg of the the extract of *Hypericum perforatum* or 3 x 25 mg maprotiline tablets of identical appearance. Effectiveness was determined using the Hamilton Depression Scale (HAMD), the Depression Scale according to von Zerssen (D-S), and the Clinical Global Impression Scale (CGI).

The total score of the HAMD scale dropped during the 4 weeks of therapy in both treatment groups by about 50%. The mean values of the D-S scale and the CGI scale showed similar results, and after 4 weeks of therapy, no significant differences in either treatment group were noticed. The onsets of the effects occurred up to the second week of treatment, but were observed earlier with maprotiline than with the hypericum extract. On the other hand, maprotiline treatment resulted in more cases of tiredness, mouth dryness, and heart complaints.[10]

Hypericum extract vs imipramine

Vorbach, Hubner and Arnoldt (1994), in a double-blind study involving 135 depressed patients, investigated comparative effect of *Hypericum perforatum* and imipramine. The inclusion criteria included typical depression history including single episode (296.2), several episodes (296.3), depressive neurosis (300.4), and adjustment disorder with depressed mood (309.0) in accordance with DSM-III-R. The dosage was 3 x 300 mg extract of *Hypericum perforatum* or 3 x 25 mg imipramine daily. The duration of the treatment was 6 weeks. Main assessment criteria were the Hamilton Depression Scale (HAMD), the Depression Scale according to von Zerssen (D-S) and the Clinical Global Impressions (CGI). In both treatment groups, a parallel reduction of the Hamilton score from 20.2 to 8.8 (LI 160, n = 67) or from 19.4 to 10.7 (imipramine, n = 68), and the transformed D-S point values from 39.6 to 27.2 (LI 160) and 39.0 to 29.2 (imipramine) were found. The analysis of Clinical Global Impressions (CGI) revealed comparable results in both treatment groups. In the group receiving extract of *Hypericum perforatum* (LI 160) fewer side effects were found as compared to imipramine.[11]

More investigations

Hansgen, Vesper and Ploch (1994) in a double-blind study investigated efficacy of extract of *Hypericum perforatum* or with placebo, in seventy-two depressive patients. The duration of the treatment was 6 weeks. Inclusion criterion was a major depression in accordance with DSM-III-R. The changes were assessed using four psychometric scales (HAMD, D-S, BEB, CGI). After 4 weeks of therapy, the statistical evaluation revealed a significant improvement in all four psychometric tests in the active group as compared to the placebo group. After switching the placebo group to active treatment (5th to 6th week of therapy), significant improvements were found in the original placebo group. No serious side effects were observed.[12]

Hubner, Lande and Podzuweit (1994) in a randomized, placebo-controlled, double-blind study, investigated effect of extract of *Hypericum perforatum* in 39 patients with depression with somatic symptoms. The therapy lasted for 4 weeks; the dosage was 300 mg three times daily. At the onset of the study as well as after 2 and 4 weeks, the following criteria were analyzed: HAMD, B-L, CGI, and vegetative symptoms. The results show a significant improvement in the active treatment group at the 5% level as compared to placebo. Seventy percent of the patients treated with extract of *Hypericum perforatum* were free of symptoms after 4 weeks. Typical symptoms of the depression such as lack of activity, tiredness, fatigue, and disturbed sleep, responded well to extract of *Hypericum perforatum*. No undesirable side effects were observed.[13]

Hypericum perforatum prevents reuptake of serotonin

Perovic and Muller (1995) reported that the methanolic extract of Hypericum lacks protective effect against N-methyl-D-aspartate (NMDA-) or gp120- (from the HIV virus) induced cytotoxicity. The study however established that Hypericum extract causes a 50% inhibition (IC50 value) of serotonin uptake by rat synaptosomes at a concentration of 6.2 microglml. The study threw some light on antidepressant action of *Hypericum perforatum*. The authors concluded that the antidepressant activity of Hypericum extract is due to an inhibition of serotonin uptake by postsynaptic receptors.[14]

Witte *et al.* (1995) in a multicenter, placebo-controlled double-blind trial, studied the effect of 100 to 120 mg of the hypericum preparation (100 to 120 mg, twice a day) in 97 patients. The efficacy of the treatment was assessed in terms of Hamilton Depression Scale (HAMD), the von Zerssen Depressivity Scale (D-S) and the Clinical Global Impression Scale (CGIS). After complete study, the researchers reported significant improvement in the symptoms of depression. More so hypericum demonstrated an anxiolytic effect. No side effect was recorded.[15]

Hypericum and seasonal affective disorder (SAD)

Kasper (1997) studied the efficacy of *Hypericum perforatum* in seasonal affective disorder (SAD). The study, which lasted for 4 weeks, recorded a significant reduction in the total score of the Hamilton Depression Rating Scale using 900 mg of hypericum. There was no significant difference when bright light therapy was combined with hypericum, compared to the situation without bright light therapy. Overall, hypericum was well tolerated and therefore the data suggest that pharmacological treatment with hypericum may be an efficient therapy in patients with SAD, which needs to be substantiated in further controlled studies.[16]

Biochemical investigation of antidepressant activity of Hypericum perforatum

Muller, Rolli, Schafer and Hafner (1997) tested possible effects of standardized extract of *Hypericum perforatum* in several biochemical models relevant for the mechanism of action of other antidepressant drugs. While the standardized extract of *Hypericum perforatum* demonstrated a weak MAO-A and MAO-B inhibiting activity, but inhibited the synaptosomal uptake of principle neurotransmitters (serotonin, dopamine and norepinephrine) with about equal affinity-2 micrograms/ml). Sub chronic treatment of rats with hypericum extract led to a significant down-regulation of beta-receptors and to a significant up-regulation of 5-HT2-receptors in the frontal cortex. The researchers concluded that antidepressant action of Hypericum extract involves same biochemical mechanisms as that of standard antidepressants.[17]

Comparative study of two Hypericum perforatum extracts (difference in hyperforin content)

Laakmann, Schule, Baghai and Kieser (1997) in a randomized, double-blind, placebo-controlled, multicenter study, investigated the clinical efficacy and safety of two different extracts of *Hypericum perforatum*. The manufacturing process for the two Hypericum preparations was identical, so that they differed only in their hyperforin content. The study included 147 male and female out patients suffering from mild or moderate depression according to DSM-IV criteria. Following a placebo run-in period of three to seven days, the patients were randomized to one of three treatment groups: During the 42-day treatment period, they received 3 x 1 tablets of either placebo, *Hypericum perforatum* extract (300 mg, with a content of 0.5% hyperforin), or *Hypericum perforatum* extract (300 mg, with a content of 5% hyperforin).

Efficacy regarding depressive symptoms was assessed on days 0, 7, 14, 28, and 42, using the Hamilton Rating Scale for Depression (HAMD, 17-item version) and the Depression Self-Rating Scale (D-S) according to von Zerssen. In addition, the severity of illness was also rated by the investigators on days 0 and 42 using the Clinical Global Impression (CGI) scale. The last observation of patients withdrawn from the trial prematurely was carried forward. At the end of the treatment period (day 42), the patients receiving WS 5572 (5% hyperforin) exhibited the largest HAMD reduction versus day 0 (10.3 ± 4.6 points; mean ± SD), followed by the WS 5573 group (0.5% hyperforin; HAMD reduction 8.5 ± 6.1 points) and the placebo group (7.9 ± 5.2 points). As regards the change in the HAMD total score between day 0 and treatment end and its relationship to the hyperforin dose, a significant monotonic trend was demonstrated in the Jonckheere-Terpstra test ($p = 0.017$). In pair wise comparisons, WS 5572 (5% hyperforin) was superior to placebo in alleviating depressive symptoms according to HAMD reduction (Mann-Whitney U-test: $p = 0.004$), whereas the clinical effects of WS 5573 (0.5% hyperforin) and placebo were descriptively comparable. These results show that the therapeutic effect of *Hypericum perforatum* in mild to moderate depression depends on its hyperforin content.[18]

Comparison of antidepressant activity of Indian *Hypericum perforatum* vs *imiparmaine*

Kumar, Singh, Jaiswal and Bhattacharya (1999) investigated antidepressant activity of a standardized 50% aqueous ethanolic extract of Indian *Hypericum perforatum*. Various experimental tests (behavioral despair, learned helplessness, tail suspension and reserpine-induced hypothermia) were studied in animals. Studies indicated that single dose administration of Indian *Hypericum perforatum* had very little or no acute behavioral effects, hence the Indian *Hypericum perforatum* was administered orally at two dose levels (100 and 200 mg/kg, p.o.) once daily for three consecutive days, while imipramine (15 mg/kg, i.p.), a clinically used antidepressant agent, was administered acutely to rats and mice of either sex as the standard drug. Controls animals were treated similarly with equal volume of vehicle (0.3% carboxymethyl cellulose). The researchers concluded that Indian *Hypericum perforatum* extract demonstrated significant antidepressant activity. Thus Indian *Hypericum perforatum* and imipramine treatments significantly reduced the immobility time in behavioral despair and tail suspension tests. Significant reduction in escape failures was also observed in learned helplessness test. In reserpine-induced hypothermia test Indian *Hypericum perforatum* and imipramine inhibited reserpine induced hypothermia in a dose dependent manner. It was concluded that the observed antidepressant activity of Indian *Hypericum perforatum* was comparable to imipramine.[19]

Inhibition of the uptake principle neurotransmitters by Hypercium perforatum

Wonnermann, Singer and Muller (1999) explained a new mechanism for antidepressant action of *Hypercium perforatum*. The researchers in their work on standardized extract of *Hypericum perforatum* have demonstrated that the extract not only inhibits the uptake of serotonin, nor epinephrine, dopamine, but also inhibits gamma amino butyric acid (GABA) and L-glutamate. To investigate the possible role of hyperforin in antidepressant activity of Hypericum extract, the uptake inhibition of hyperforin was compared with that of Na^+-ionosphere monoensin and ouabain. In the presence of amiloride and MIA (non-selective inhibitors of Na^+-H^+ antiporters and Na^+-

channels), benzamil (Na^+ channel blocker), and EIPA (inhibitor of Na^+- H^+ exchangers), the researchers proved reduced gamma amino butyric acid (GABA) and L-glutamate by hyperforin. On the other hand, monoesin or ouabain inhibition was not affected. The researchers concluded that hyperforin inhibits serotonin uptake by elevating free intracellular Na^+.[20]

Anxiolytic activity of Indian Hypericum perforatum

Kumar, Jaiswal, Singh and Bhattacharya (2000) investigated the anxiolytic activity of 50% ethanolic extract of Indian *Hypericum perforatum* in rats. The work included various experimental paradigms of anxiety viz. open field exploratory behavior, elevated plus maze, elevated zero maze, novelty induced suppressed feeding latency and social interaction tests. Pilot studies indicated that single dose administration of Indian *Hypericum perforatum* had little to no acute behavioral effects, hence the extract of Indian *Hypericum perforatum* was administered orally at different dose levels once daily for three consecutive days, while lorazepam (0.5 mg/kg, i.p.) was administered acutely. Indian *Hypericum perforatum* extract (100 and 200 mg/kg, p.o.) showed significant anxiolytic effects on all the paradigms of anxiety.

The results indicate that Indian *Hypericum perforatum* and lorazepam induced a significant increase in open field ambulation and slight increase in rearings and activity in centre, whereas grooming and fecal droppings remain unchanged. In elevated plus maze, significant augmentation of open arm entries, open arm/ closed arm entries ratio and time spent on open arms was noted in Indian *Hypericum perforatum* treated rats. In elevated zero maze test, significant increase in time spent on open arms and entries in open arms were observed, whereas slight increase in head dips and stretched attend postures were also observed. Indian *Hypericum perforatum* and lorazepam significantly attenuated the novelty induced increase in feeding latency. Indian *Hypericum perforatum* treated rats also showed significant increase in social interaction in the novel environment. The Indian *Hypericum perforatum* extracts showed consistent and significant anxiolytic activity in all the tests. The effects induced by 50% ethanolic extract of Indian *Hypericum perforatum* were less marked than those of lorazepam.[21]

Nootropic activity of Indian Hypericum perforatum

Kumar, Singh, Muruganandam and Bhattacharya (2000) studied the nootropic activity of a standardized 50% ethanolic extract of Indian *Hypericum perforatum*. The study included experimental paradigms of learning and memory (transfer latency on elevated plus-maze, passive avoidance, active avoidance, scopolamine and sodium nitrite induced amnesia. Work indicated that single dose administration of Indian *Hypericum perforatum* had little or no acute behavioral effects, hence the extract of Indian *Hypericum perforatum* was administered orally at two dose levels (100 and 200 mg/kg, p.o.), once in daily for three consecutive days, while piracetam (500/kg, i.p.), a clinically used nootropic agent, was administered acutely to rats as the standard drug.

Control rats were treated with equal volume of vehicle (0.3% carboxymethyl cellulose). Indian *Hypericum perforatum* and piracetam when given alone shortened the transfer latency on day 1, 2, 9 and also antagonized the amnesic effects of scopolamine and sodium nitrite on the transfer latency significantly. Indian *Hypericum perforatum* had no significant per se effect on the retention of the passive avoidance in rats. Only the higher dose (200 mg/kg, p.o.) produced a significant reversal of scopolamine induced passive avoidance retention deficit but no significant reversal was observed with sodium nitrite. Piracetam showed significant per se facilitatory effect on passive avoidance retention and also reversed the scopolamine and sodium nitrite induced impaired PA retention.

In the AA test, Indian *Hypericum perforatum* in both the doses, and piracetam, facilitated the acquisition and retention of active avoidance in rats and the Indian *Hypericum perforatum* effects were found to be dose dependent. Both the doses of Indian *Hypericum perforatum* and piracetam significantly attenuated the scopolamine and sodium nitrite induced impaired retention of active avoidance. These results indicate a possible nootropic action of Indian *Hypericum perforatum*, which was qualitatively comparable with that induced by piracetam.[22]

Hypericum vs. fluoxetine

Schrader (2000) in a in a randomized, double-blind, investigated comparative antidepressant effect of *Hypericum perforatum* extract tablets and the selective serotonin reuptake inhibitor, fluoxetine in patients with mild-moderate depression. The age group included 16-24 and number of patients was 240 (fluoxetine: 114 (48%) and hypericum: 126 (52%). After 6 weeks' treatment, mean HAM-D at endpoint decreased to 11.54 on hypericum and to 12.20 on fluoxetine ($P < 0.09$), while mean Clinical Global Impression (CGI) item I (severity) was significantly ($P < 0.03$) superior on hypericum, as was the responder rate ($P = 0.005$). *Hypericum perforatum* safety was substantially superior to fluoxetine, with the incidence of adverse events being 23% on fluoxetine and 8% on hypericum. Side effects recorded in fluoxetine group included agitation, gastrointestinal disturbances, retching, dizziness, tiredness, nervousness and erectile dysfunction. The group receiving *Hypericum perforatum* recorded gastrointestinal disturbances as major side effect. The researchers concluded that both *Hypericum perforatum* and fluoxetine were having equally potent antidepressant activity. *Hypericum perforatum* was superior to fluoxetine in improving the responder rate and in terms of overall incidence of side-effects.[23]

Work on long-term administration of Hypericum

Butterweck *et al.* (1994) investigated the long-term effect of *Hypericum perforatum* and hypericin on monoamine oxidase levels in rat brain (hypothalamus and hippocampus). The researchers used a High Performance Liquid Chromatography to investigate the short-term and long term administration of tricyclic antidepressant imipramine, extract of Hypericum perforatum and hypericin on levels of neurotransmitters serotonin (5-hydroxytrpytmaine), catecholamines (nor epinephrine and dopamine). Hypothalamus and hippocampus were selected because these areas are target of antidepressant drug action. Imipramine, extract of *Hypericum perforatum* and hypericin (given daily for 8 weeks) significantly increased the serotonin levels in the hypothalamus. However levels of 5-hydroxyindoleacetic acid were not altered in hypothalamus and hippocampus. The ratio of serotonin/ 5-hydroxyindoleacetic acid ratio was lowered in the hypothalamus and hippocampus after 8 weeks of treatment. In comparison to *Hypericum perforatum*, hypericin and imipramine decreased the dopamine and 5-hydroxyindoleacetic acid levels in the hypothalamus. The dopamine turnover ratio in hypothalamus was significantly reduced after 8 weeks by extract of Hypericum perforatum and hypericin and slightly by imiparmine. Hypericin reduced the nor epinephrine concentration in the hippocampus after 2 and 8 weeks. Extract of *Hypericum perforatum* and imipramine shoed similar result after 8 weeks. The researchers concluded that only long-tem administration of *Hypericum perforatum* and hypericin modify the levels of neurotransmitters in the brain regions involve din the pathophysiology of depression.

Hypericum and obsessive-compulsive disorder (OCD)

Taylor and Kobak (2001) investigated efficacy of *Hypericum perforatum* in the treatment of obsessive-compulsive disorder. Twelve subjects were evaluated with a primary DSM-IV diagnosis of OCD of at least 12 months' duration. Treatment lasted for 12 weeks, with a fixed dose of 450 mg of 0.3% hypericin (a psychoactive compound in Hypericum) twice daily (extended-release formulation). Weekly evaluations were conducted with the Yale-Brown Obsessive Compulsive Scale, the Patient Global Impressions of Improvement Scale and the Clinical Global Impressions of Improvement scale and monthly evaluation with the Hamilton Rating Scale for Depression. A significant change from baseline to endpoint was found, with a mean Yale-Brown Obsessive Compulsive Scale change of 7.4 points ($p = 0.001$). Significant change occurred at 1 week ($p = 0.020$) and continued to increase throughout the trial. At endpoint, 5 of 12 were rated "much" or "very much improved" on the clinician-rated Clinical Global Impressions of Improvement scale, 6 were "minimally improved," and 1 had "no change." The most common side effects reported were diarrhea and restless sleep. Significant improvement was found with Hypericum, with a drop in Yale-Brown Obsessive Compulsive Scale score similar to that found in clinical trials. The researchers stressed the need of a placebo-controlled study of Hypericum in obsessive-compulsive disorder.[24]

Mode of action of hyperforin

1. Muller, Singer and Wonnemann (2001) suggested that hyperforin is the major antidepressive constituent of *Hypericum perforatum* extract. It not only inhibits the neuronal uptake of serotonin, norepinephrine and dopamine like many other antidepressants, but also inhibits

gamma-amino butyric acid (GABA) and L-glutamate uptake. This broad-spectrum effect is obtained by an elevation of the intracellular Na^+ concentration. It is probably due to activation of sodium conductive pathways (not yet established but most likely ionic channels). The researchers claim hyperforin as first member of a new class of compounds with a preclinical antidepressant profile having a novel and distinct mechanism of action.[25]

2. Eckert and Muller (2001) claimed that hyperforin from *Hypericum perforatum* affects several ionic conductance mechanisms in brain cells although the mechanism of action is still is awaited. The researchers tested the effects of hyperforin on the fluidity of crude brain membranes from young guinea pigs. Fluidity measurements were performed with three different fluorescent probes. Diphenylhexatriene and trimethylam-monium-diphenylhexatriene anisotropy measurements were inversely correlated with the flexibility of fatty acids in the membrane hydrocarbon core and in the hydrophilic area of membrane phospholipids, respectively. The ratio of pyrene excimer to monomer fluorescence intensities was used as an indicator of membrane annular and bulk fluidity. Incubation of brain membranes with relatively high concentrations of hyperforin sodium salt (10 micromol/l) resulted in increased Diphenylhexatriene and decreased trimethylammonium-diphenylhexatriene anisotropy, respectively, indicating that hyperforin modifies specific membrane structures in different ways. It decreases the flexibility of fatty acids in the membrane hydrocarbon core, but fluidizes the hydrophilic region of membrane phospholipids. Interestingly, relatively low concentrations of hyperforin (0.3 micromol/l) significantly decreased the annular fluidity of lipids close to membrane proteins. These findings are remarkable, as inhibition of several neurotransmitter-uptake systems and modulation of several ionic conductance mechanisms by hyperforin occur in the same concentration range. However, bulk fluidity was unchanged by this low hyperforin concentration. The results warrant a physicochemical interaction of hyperforin with specific membrane structures, which

probably contribute to its novel pharmacological properties.[26]

Antidepressant activity of Indian Hypericum perforatum

Kumar, Singh and Bhattacharya (2001) studied the effect of acute administration of 50% standardized ethanolic extract of Indian *Hypericum perforatum* on the rat brain concentrations of monoamines and their metabolites in five different brain regions, (hypothalamus, hippocampus, striatum, pons-medulla and frontal cortex) by a HPLC technique. Indian *Hypericum perforatum* extract was administered at the doses of 50 and 200 mg/kg, p.o. and the brain monoamines were assayed after 30 min of the treatment. Indian *Hypericum perforatum* treatment resulted in significant decreased in the levels of serotonin and its metabolite 5-hydroxy indole acetic acid and serotonin turnover in all the brain regions assayed. On the other hand, Indian *Hypericum perforatum* treatment significantly augmented the levels of norepinephrine and its metabolite methylhydroxy phenyl glycol and norepinephrine turnover in all the brain regions studied. Similarly, the levels of dopamine were also significantly augmented in the hypothalamus, striatum and frontal cortex. Likewise, the levels of dihydroxyphenyl acetic acid, the major metabolite of dopamine, were also increased in these brain areas. The researchers conclude that pharmacological studies with Indian *Hypericum perforatum* extract demonstrated two major behavioral actions, (anxiolytic and antidepressant). Further the work proves that neither reduced serotonin activity associated with anxiolytic and increased nor epinephrine and dopamine activity being associated with antidepressant activity.[27]

Cognitive activity of Hypericum extract and hyperforin

Klusa *et al.* (2001) evaluated effect of *Hypercium perforatum* extract and sodium salt of hyperforin in rat avoidance test. In a conditioned avoidance response test on rats, oral daily administration of hyperforin (1.25 mg/kg/day) or of the extract of *Hypericum perforatum* (50 mg/kg/ day) before the training sessions significantly improved learning ability from second day onward until the day seven. The memory attained during seven days of treatment was retained for nine days without further treatment. In a passive avoidance response test, a single oral dose of hyperforin not only resulted in improved memory acquisition but completely reversed scopolamine induced amnesia. Single dose of *Hypericum perforatum* extract (50 mg/kg / day) did not produce significant effect in passive avoidance test in the animals.[28]

Pharmacology of Hyperforin acetate

Zanoli, Rivasi, Baraldi and Baraldi (2002) investigated pharmacological activity of hyperforin acetate. In the forced swimming test, triple administration of hyperforin resulted in significant reduction in immobility time of rats. In the light-dark test, hyperforin acetate demonstrated anxiolytic effect but smaller as compared to diazepam. More so hyperforin acetate demonstrated muscle relaxant activity as evident from reduced locomotion in rats. Acute as well as repeated doses of hyperforin acetate did not altered pentobarbital sleeping time in rats.[29]

Pseudohypericin as antidepressant constituent

Simmen *et al.* (2003) investigated the possible role of pseudohypericin in antidepressant action of *Hypericum perforatum.* Earlier *in vivo* studies have shown that hypericin reduces the activity of the hypothalamic-pituitary-adrenal axis. It has been suggested that exacerbated hypothalamic-pituitary-adrenal axis activity has been associated with the incidence of depression. Corticotropin-releasing factor is thought to be major determinant in the regulation of the hypothalamic-pituitary-adrenal activity via activation of CRF (1) receptors. The researchers investigated CRF (1) receptors antagonist activity of hyperforin, hypericin and pseudohypericin. They measured the effects of hyperforin, hypericin and pseudohypericin on CRF-stimulated cAMP formation in the recombinant Chinese hamster ovary cells. Hyperforin, hypericin and pseudohypericin were also tested against calcitonin in some cells. Only pseudohypericin demonstrated anatogomstic activity on CRF. Hypericin and hyperforin affected both the CRF and calcitonin. It was conclude that pseudohypericin was the active CRF antagonist and thus antidepressant activity of *Hypericum perforatum* is not only due to hypericin or hyperforin but pseudohypericin also contributes to the antidepressant mechanism.[30]

Hypericum vs. amitriptyline

Wheatley (1997) performed a controlled, randomized, multi-centre trial on antidepressant activities of *Hypericum perforatum* (LI 160; 900 mg daily in divided doses) and well known tricyclic antidepressant, amitriptyline (75 mg). The trial lasted for 6-weeks. The observations based on the Hamilton Depression Scale response rate, exhibited no statistically significant difference between the groups, although response was better in amitriptyline group. The secondary efficacy parameters, decreases in the total Hamilton Depression and Montgomery-Asberg scores, showed a significant advantage for amitriptyline, but only at week 6. *Hypericum perforatum* was well-tolerated as compared to amitrip-tyline, In *Hypericum perforatum* group, 37% patients reported adverse events, as compared to 64% in the amitriptyline group.[31]

Hypericin- interaction with cholinergic and σ receptors

Raffa (1998) in a study has shown that hypericin (1.0 μM) has modest affinity for muscarinic cholinergic receptors (49%) as well as σ receptors (48%). σ receptors are located in areas of the brain involved with the regulation of emotions, such as the limbic system, and a role for σ receptors in the antidepressive action of synthetic antidepressants has been postulated.[32]

Xanthones as antidepressant constituent

Muruganandam *et al.*, (2000) studied the possible role of xanthones in the antidepressant activity of *Hypericum perforatum*. Commercial extract were found to contain relatively small amounts of xanthones (1-3%) whereas an Indian specimen contained 2-4% xanthones. A xanthone-enriched fraction from Indian specimen administered to rats (5 mg/kg p.o.) showed significant activity in the forced swimming test and exhibited receptor binding inhibitory activities in the frontal cortex of the rats, suggesting upregulation benzodiazepine and serotonin receptors and downregulation of dopamine D2 receptors. Four trioxygenated xanthones isolated from the plant also exhibited significant antidepressant activity. A hyperforin-enriched fraction at the same dosage was equipotent to the xanthone-enriched fraction. When the two extracts were combined in suboptimal dosages of 2.5 mg each, they caused a significant antidepressant effect which was slightly greater in extent than exhibited by either the hyperforin-enriched- or the xanthone-enrichedfraction.[33]

Flavonoids as antidepressant constituent

Butterweck *et al.* (2000) demonstrated that a flavonoid fraction obtained from crude *Hypericum perforatum* was active in forced swimming test. The fraction was further separated into two sub fractions. Both the sub fractions were active in forced swimming test within varied doses. Further work on first sub fraction showed that main constituents were hyperoside, isoquercitrin, miquelianin and quercetin. Second sub fraction contained small amounts of hyperoside and astiblin. Further testing on these compounds showed that except quercetin all the all were active in the forced swimming test.[34]

Hypericum perforatum and prolactin

Franklin and Cowen (2001) in their work demonstrated that *Hypericum perfortaum* may lower serum prolactin. They studied the effect of acute and sub-chronic treatment of a commercial preparation based on methanolic extract of *Hypericum perforatum* on plasma hormones and brain neurotransmitters. The study was performed in rats and health humans. The study also included the effects of hyperforin and hypericin with respect to the extract. In the acute study, the normal subjects were given nine tablets of the commercial preparation of *Hypercium perforatum* and placebo. The normal subjects showed significant increase in salivary cortisol and plasma growth hormone and decrease in plasma prolactin with respect to placebo. Also plasma hormone levels were associated with a rise in plasma hyperforin.

In animals, acute treatment with *Hypericum perforatum*, hyperforin and hypericin caused significant increase in plasma corticosterone and brain cortical serotonin. The corticosterone response was affected by ketanserin (5-HT2 antagonist) but not by WAY-100635 (5-HT1A antagonist).

When the sub-chronic and acute treatment were compared (two dosed of *Hypericum perforatum* extract were used), it was observed that plasma corticosterone level was significantly reduced. More so plasma prolactin was significantly increased by hyperforin and the extract but not by hypericin. Further, it was associated with rise in brain cortical dopamine. The authors concluded that *Hypericum perforatum* may effect plasma hormonal changes through serotonin and dopamine mediated processes. More so hyperforin seems to be more active as compared to hypericin following acute treatment.[35]

Effect of Hypericum on dopamine release

Zhou *et al* studied the possible mechanism of antidepressant activity of *Hypericum perforatum*. The researchers used *in vivo* micro dialysis to investigate effect of *Hypericum perforatum* on dopamine, 3, 4-dihydroxyphenylacetic acid, 4-hydroxy-3- methoxy-phenylacetic acid and 5-hydroxyindolacetic acid release in rat striatum in active rats. Oral administration of the extract (25mg/kg) resulted in significant rise in extra cellular level of dopamine in rat striatum. During the study, the levels of, 4-hydroxy-3-methoxy-phenylacetic acid and 5-hydroxyindolacetic acid remained similar to those of control rats. The researchers concluded that *Hypericum perforatum* extract causes increased dopamine release and delays decrease of 3, 4-dihydroxyphenylacetic acid.[36]

Hypericum perforatum and alcoholism

Rezvani *et al.* (1999) studied the effect of *Hypericum perforatum* extract on voluntary alcohol intake in two different genetic animal models of human alcoholism: fawn-hooded (FH) and high-alcohol drinking (HAD) rats. The rats received a single oral administration (5 ml/kg) of either vehicle or one of the five doses (100, 200, 400, 600, and 800 mg/kg) of *Hypericum perforatum* extract. The oral administration of *Hypericum perforatum* extract significantly reduced alcohol intake in both fawn-hooded and high-alcohol drinking rats. In a third study, fawn-hooded rats did not develop tolerance to the suppressant effects of *Hypericum perforatum* on alcohol intake and preference following oral administration of (400 mg/kg) of the extract for 15 consecutive days. The observations suggest that *Hypericum perforatum* extract should be evaluated clinically as a potential therapeutic agent in the treatment of alcoholism.[37]

Hypericum perforatum and sertraline

The efficacy and safety of *Hypericum perforatum* extract for the treatment of major depression was evaluated in a double-blind, randomized, placebo-controlled trial. 340 patients with definite history of major repression were recruited for the trail. Patients were randomly assigned to receive *Hypericum perforatum* extract, placebo, or sertraline (as an active comparator) for 8 weeks. The daily dose of *Hypericum perforatum* range from 900 to 1500 mg and that of sertraline from 50 to 100 mg. Based on the HAM-D and Clinical Global Impressions scores neither sertraline nor *Hypericum perforatum* were significantly different from placebo. The random regression parameter estimate for mean (SE) change in HAM-D total score from baseline to week 8 (with a greater decline indicating more improvement) was -9.20 (0.67) (95% confidence interval [CI], -10.51 to -7.89) for placebo versus -8.68 (0.68) (95% CI, -10.01 to -7.35) for H perforatum (P =0.59) and -10.53 (0.72) (95% CI, -11.94 to -9.12) for sertraline (P =0.18). Full response occurred in 31.9% of the placebo-treated patients vs 23.9% of the H perforatum-treated patients (P =0.21) and 24.8% of sertraline-treated patients (P =0.26). Adverse-effect profiles for *Hypericum perforatum* and sertraline differed relative to placebo. It was concluded that efficacy of *Hypericum perforatum* in severe major depression was not justified[38].

References

1. Muldner, H. and Zoller, M. (1984). Antidepressive effect of a Hypericum extract standardized to an active hypericine complex. *Arzneimittel forschung* 34 (8): 918-920.
2. Okpanyi, S.N. and Weischer, M.L. (1987). Animal experiments on the psychotropic action of a Hypericum extract. *Arzneimittelforschung* 37 (1): 10-13.
3. Schmidt, U. and Sommer, H. (1993). St. John's wort extract in the ambulatory therapy of depression. Attention and reaction ability are preserved. *Fortschr Med* 111 (19): 339-342.
4. Sommer, H. and Harrer, G. (1994). Placebo-controlled double-blind study examining the effectiveness of a hypericum preparation in 105 mildly depressed patients. *J Geriatr Psychiatry Neurol* 7 (1): 9-11.

5. Muller, W.E. and Rossol, R. (1994). Effects of hypericum extract on the expression of serotonin receptors. *J Geriatr Psychiatry Neurol* 7 (1): 63-64.
6. Thiele, B. and Ploch, M. (1994) Modulation of cytokine expression by hypericum extract. *J Geriatr Psychiatry Neurol* 7 (1): 60-62.
7. Bladt S. and Wagner, H. (1994). Inhibition of MAO by fractions and constituents of hypericum extract. *J Geriatr Psychiatry Neurol* 7 (1): 57-59.
8. Thiede, H.M. and Walper, A. (1994) Inhibition of MAO and COMT by hypericum extracts and hypericin. *J Geriatr Psychiatry Neurol* 7 (1): 54-56.
9. Johnson, D. Ksciuk, H. and Woelk, H, (1994). Sauerwein-Giese E and Frauendorf A. Effects of hypericum extract LI 160 compared with maprotiline on resting EEG and evoked potentials in 24 volunteers. *J Geriatr Psychiatry Neurol* 7 (1): 44-46.
10. Harrer, G. Hubner, W.D. and Podzuweit, H. (1994). Effectiveness and tolerance of the hypericum extract LI 160 compared to maprotiline: a multicenter double-blind study. *J Geriatr Psychiatry Neurol* 7 (1): 24-28.
11. Vorbach, E.U. Hubner, W.D. and Arnoldt, K.H. (1994). Effectiveness and tolerance of the hypericum extract LI 160 in comparison with imipramine: randomized double-blind study with 135 outpatients. *J Geriatr Psychiatry Neurol* 7 (1):19-23.
12. Hansgen, K.D. Vesper, J. and Ploch, M. (1994). Multicenter double-blind study examining the antidepressant effectiveness of the hypericum extracts LI 160. *J Geriatr Psychiatry Neurol* 7 (1): 15-18.
13. Hubnor, W.D. Lande, S and Podzuweit, H. (1994). Hypericum treatment of mild depressions with somatic symptoms. *J. Geriatr Psychiatry Neurol* 7 (1): 12-14.
14. Perovic, S. and Muller, W.E. (1995). Pharmacological profile of hypericum extract. Effect on serotonin uptake by postsynaptic receptors. *Arzneimittelforschung* 45 (11):1145-1148.
15. Witte, B., Harrer, G., Kaptan, T., Podzuweit, H. and Schmidt, U. (1995). Treatment of depressive symptoms with a high concentration Hypericum preparation. A multicenter placebo-controlled double-blind study. *Fortschr Med* 113 (28): 404-448.
16. Kasper, S. (1997). Treatment of Seasonal Affective Disorder (SAD) with Hypericum extract. *Pharmacopsychiatry* 30 (2): 89–93.

17. Muller, W.E., Rolli, M., Schafer, C. and Hafner, U. (1997). Effects of Hypericum extract (LI 160) in Biochemical Models of Antidepressant Activity. *Pharmacopsychiatry* 30 (2): 102–107.

18. Laakmann, G., Schule, C., Baghai, T. and Kieser, M. (1998) St. John's wort in mild to moderate depression: the relevance of hyperforin for the clinical efficacy. *Pharmacopsychiatry* 31 (1): 54-59.

19. Kumar, V., Singh, P.N., Jaiswal, A.K. and Bhattacharya, S.K. (1999). Antidepressant activity of Indian *Hypericum perforatum* Linn in rodents. *Indian J Exp Biol* 37 (12): 1171-1176.

20. Singer, A., Wonnemann, M. and Muller, W.E. (1999). Hyperforin, a major antidepressant constituent of St. John's wort, inhibits serotonin uptake by elevating free intracellular Na^+. *Pharmacology and Experimental Therapeutics* 290 (3): 1363-1368.

21. Kumar, V., Singh, P.N., Jaiswal, A.K. and Bhattacharya, S.K. (1999). Antidepressant activity of Indian *Hypericum perforatum* Linn in rodents. *Indian J Exp Biol* 37 (12): 1171-1176.

22. Kumar, V., Jaiswal, A.K., Singh, P.N. and Bhattacharya, S.K. (2000). Anxiolytic activity of Indian *Hypericum perforatum* Linn: an experimental study. *Indian J Exp Biol* 38 (1): 36-41.

23. Schrader, E. (2000). Equivalence of St John's wort extract (Ze 117) and fluoxetine: a randomized, controlled study in mild-moderate depression. *Int Clin Psychopharmacol* 15 (2): 61-68.

24. Taylor, L.H. and Kobak, K.A. (2000). An open-label trial of St. John's Wort (Hypericum perforatum) in obsessive-compulsive disorder. *J Clin Psychiatry* 61: 575-578.

25. Muller, W.E. Singer, A. and Wonnemann, M. (2001). Hyperforin-antide-pressant activity by a novel mechanism of action. *Pharmacopsychiatry* 34 (1): 98-102.

26. Eckert, G.P. and Muller, W.E. (2001). Effects of hyperforin on the fluidity of brain membranes. *Pharmacopsychiatry* 34 (1): 22-25.

27. Kumar, V., Singh, P.N. and Bhattacharya, S.K. (2001). Neurochemical studies on Indian *Hypericum perforatum* L. *Indian J Exp Biol.* 39 (4): 334-338.

28. Klusa, V., Germane, S., Noldner, M. and Chatterjee, S.S. (2001). Hypericum extract and hyperforin: memory-enhancing properties in rodents. *Pharmacopsychiatry* 34 (1): 61-69.

29. Zanoli, P., Rivasi, M., Baraldi, C. and Baraldi, M. (2002). Pharmacological activity of hyperforin acetate in rats. *Behav Pharmacol.* 13 (8): 654-651.

30. Simmen, U., Bobirnac, I., Ullmer, C., Lubbert, H., Berger Buter, K., Schaffner, W. and Schoeffter, P. (2003) Antagonist effect of pseudohypericin at CRF1 receptors. *Eur J Pharmacol* 548 (3): 251-256.

31. Wheatley, D. (1997). LI 160, An Extract of St. John's wort, Versus Amitriptyline in Mildly to Moderately Depressed Outpatients: A Controlled 6-week Clinical Trial. *Pharmacopsychiatry* 30(27): 77–80.

32. Raffa, R.B. (1998). Screen of receptor and uptake-site activity of hypericin component of St. John's wort reveals sigma receptor binding. *Life Sciences* 62: 265-L270.

33. Muruganandam, A.V., Ghosal, S. and Bhattacharya, S.K. (2000). The role of xanthones in the antidepressant activity of *Hypericum perforatum* involving dopaminergic and serotonergic systems. *Biogenic Amines* 15: 553-567.

34. Butterweck, V. Jurgenliemk, G. Nahrstedt, A. and Winthehoff, H. (2000). Flavonoids from *Hypericum perforatum* show antidepressant activity in the forced swimming test. Thieme: Fachzeitschriften: *Planta Medica*: 1/2000: Rapid communication.

35. Franklin, M. and Cowen, PJ. (2001). Researching the antidepressant actions of *Hypericum perforatum* in animals and man. *Pharmacopsychiatry* 34 (1): 29-37.

36. Zhou, Q., Gitzen, J., Cregor, M., Sotoyama, H., Zhu, Y., Janle, E. and Kissinger, P. (2000). Effect of acute administration of *Hypercium perforatum* extract on dopamine release on the rat brain. *Bioanalytical Systems*, Inc., 2701 Kent Avenue, West Lafytee, IN, 47906.

37. Rezvani, A.H., Overstreet, D.H., Yang, Y. and Calrk, E. (1999). Attenuation of alcohol intake by extract of *Hypericum perforatum* (St. John's wort) in two different strains of alcohol-preferring rats. *Alcohol and Alcoholism* 34 (5): 699-705.

38. No author listed (2002). Effect of *Hypericum perforatum* (St John's wort) in major depressive disorder: a randomized controlled trial. *JAMA* 287 (14): 1807-1814.

6

Hypericum and Digestive System

Berghofer & Holzl (1989) reported anti-ulcerogenic effect of amentoflavone.[1] Ozturk *et al.* (1992) reported hepatoprotective activity of *Hypericum perforatum*. It was assumed due to decreased barbiturate-induced sleeping time in mice, increased bile flow in cannulated rats, and decreased carbon tetrachloride-induced hepatotoxicity in rats.[2]

Krylov and Ibatov (1993) reported the use of an infusion of *Hypericum perforatum* in the combined treatment of alcoholics with peptic ulcer and chronic gastritis. *Hypericum perforatum* herbal infusion (1 glass 4-5 times daily) in association with rational psychotherapy was prescribed to 57 outpatients with alcoholism and concomitant diseases (peptic ulcer and chronic gastritis) for two months. The authors concluded that treatment with *Hypericum perforatum* coupled with rational psychotherapy was effective in relieving symptoms of peptic ulcer and chronic gastritis.[3]

References

1. Berghofer, R. and Holzl, J. (1989). Isolation of 13', II8-biapigenin (amentoflavone) from *Hypericum perforatum*. *Planta Med* 55: 91.

2. Ozturk, Y. *et al.* (1992). Hepatoprotective activity of *Hypericum perforatum* L. alcoholic extract in rodents. *Phytotherapy Res* 6: 44-46.

3. Krylov, A.A. and Ibatov, A.N. (1992). The use of an infusion of St.-John's-wort in the combined treatment of alcoholics with peptic ulcer and chronic gastritis. *Lik Sprava* (2-3): 146-8

7

Hypericum and Cancer

Thomas (1992) examined potential of hypericin as agent for chemotherapy and for photodynamic therapy of neoplastic disease. The model systems used to test for activity were a mitochondrial succinoxidase enzyme activity assay, an EMT6 mouse mammary carcinoma clonogenic assay, and an MX-1 human mammary carcinoma antitumor assay. Parallel experiments were conducted in the presence and absence of light. Pure hypericin and acetone extracts of *Hypericum perforatum* were tested for activity and yielded similar results. In the dark, hypericin inhibited succinoxidase activity in isolated mitochondrial membranes, but had no effect on EMT6 cells or MX-1 tumors. However, photoactivated hypericin exhibited activity in all test systems.

In particular, photoactivated hypericin inhibited succinoxidase activity at nmol/mg mitochondrial protein levels, EMT6 cell clonogenicity at uM levels, and MX-1 relative tumor wt growth in ip and it dosed animals at 1 mg/kg body wt levels. Experiments designed to elucidate the mechanistic parameters of hypericin's biological activity were also conducted. The results show that photosensitized inhibition of mitochondrial succinoxidase was drug dose-, light dose-, and wavelength-dependent with the greatest inhibition occurring by 600 nm light activation.

Hypericin photosensitized singlet oxygen generation was also light dose- and wavelength-dependent. Unequivocal evidence for hypericin photosensitized singlet oxygen generation was obtained using kinetic isotope ratios of products from the reaction between singlet oxygen and the germinally deuterated singlet oxygen trap, tetramethylethylene. The data are consistent with type II singlet oxygen-mediated mechanisms for hypericin photosensitized succinoxidase inhibition. The photodynamic effect of hypericin on EMT6 cells was examined under hypoxic and aerobic conditions, and photocytotoxicity was found to be oxygen-dependent.

The findings suggest that type I oxygen-independent photo-oxidation reactions do not contribute significantly to hypericin's photodynamic mechanism of action. Fluorescence photomicrographs of hypericin in EMT6 cells show that hypericin is distributed throughout the plasma and cytosolic membranes but is absent from the nucleus, indicating that the events which lead to photocytotoxicity may not involve direct damage to nuclear DNA.[1] Couldwell *et al.* (1994) investgated glioma-inhibitory activity of hypericin. Protein kinase C has been implicated in the formation and proliferation of tumors. Hypericin inhibited the growth of glioma cell lines *in vitro* and induced glioma cell death due to inhibition of Protein kinase C, as measured by [3H]-thymidine uptake. The glioma-inhibitory activity of hypericin was comparable to or greater than tamoxifen (IC50 <10 versus 10 mmol/L, respectively). The activity was enhanced by approximately 13% from exposure of hypericin to visible light.[1]

Vandenbogaerde (1995) investigated in vitro cytotoxic and in vivo antitumoral properties of photosensitized hypericin In vitro studies demonstrated a dramatic difference in sensitivity of cell lines towards photo-activated hypericin using a neutral red assay. Using confocal laser microscopy it was shown that the cytotoxicity correlated with the cellular uptake of hypericin. The in vivo antitumoral activity hypericin was tested on A431 cell xenografts in athymic nude mice. A dose-dependent antitumoral effect was observed by intraperitoneal administration of hypericin in light-treated animals; a complete inhibition of the tumor growth was achieved from 50 μg (2.5 mg/kg) hypericin ip administered. The photo-sensitizer accumulated in A431 xenografts after local light irradiation. These results show that the naphthodianthrone hypericin is a potential photochemotherapeutic agent.[2]

Alecu *et al.* (1998) examined the topical use of hypericin as a photodynamic anticancer therapy in 19 patients with skin cancer: 11 patients (median age 66) diagnosed with basal cell carcinoma and 8 patients (ages 39-72) with squamous cell carcinoma, 5 of who had not received prior treatments for their cancer. Treatment consisted of intralesional injections of hypericin followed with irradiation using visible light to photo-activate hypericin.

For ethical reasons, surgery was used after the experimental therapy to remove residual tumor lesions and to make an assessment of possible impairments caused by the phototherapy. Among the squamous cell carcinoma patients, tumors were reduced by 33% after treatment with hypericin (40-100 μg hypericin i.l. 3-5 μ/week for 2-6 weeks) and in one case the treatment appeared to cause complete remission.

One female patient showed no tumor reduction after 10 injections (540 μg hypericin) over 15 days into a tumor of the lower lip (30/20 μ10 mm in size). Analysis of the results showed that tumor reduction was dependent on initial tumor size and the dosage of hypericin administered. The analysis also revealed that in total 1,000 μg plus would be required in order for hypericin to reach acceptability as an alternative treatment for squamous cell carcinoma.

Among the patients with basal cell carcinoma, treatment with hypericin (40-200 μg hypericin intralesionally, 3-5 μ/week for 2-6 weeks) resulted in tumor reductions of less than 10% to as much as 100% in two cases with complete remission. For all 11 patients, photodynamic treatment with hypericin represented their first therapy for cancer. All these patients refused follow-up surgery and 5 months later showed no sign of the cancer remaining or recurring. In total, these patients received from 1,500□3,000 μg of hypericin. The only side effect observed was edema and erythema of a mild transient nature at the tumor site, which occurred in 3 patients in the basal cell carcinoma group and two in the squamous cell group. As the lesions healed, the researchers observed decrusting, inflammation, necrosis, and new epithelium formation, which covered the lesion. Although small, isolated groups of tumor cells remained in the dermis after healing, along with necrosis there was evidence that lymphocytes had infiltrated the remaining tumor cells. The treatment appeared to be non-toxic, tumor-selective, and left no scarring.[3]

Schempp *et al.* (2002) reported inhibition of tumour cell growth by hyperforin. The effect of hyperforin was analyzed in a range of human and rat tumour cell lines *in vivo* and *in vitro*. Hyperforin inhibited human and rat tumour cell growth (IC 50 3-

15 μM). Hyperforin resulted in dose-dependent generation of apoptotic oligonucleosomes, typical DNA–laddering and structural changes characteristic of apoptosis. In a breast cancer cell line (MT-450), hyperforin increased the activities of enzymes caspase 9 and caspase 3. However the capase inhibitor zVAD.fmk inhibited the hyperforin mediated apoptosis. In MT-450 cells, hyperforin caused damage to mitochondria. Further hyperforin caused the release of cytochrome C from isolated mitochondria. When hyperforin was compared with paclitaxel *in vivo,* both inhibited growth of MT-450 cells in rats. However hyperforin showed no signs of acute toxicity. The researchers concluded that hyperforin inhibits tumour cell growth via the activation of mitochondria-mediated apoptosis pathway.[4]

Alicia *et al.* at Tennessee State University did a comparative study of anticancer effect of *Hypercium perforatum* and *Hypercium calycinum* on cell line BT459. The anticancer effect was studied using extract concentration and cell density analyses. The amount of *Hypercium perforatum* and *Hypercium calycinum* added to each plate varied inversely with the number of recovered cells. According to researchers, the data obtained suggested that hypericin induces cell death of BT549 cancer cells.

Martarelli *et al.* (2004) reported *Hypericum perforatum* methanolic extract inhibits growth of human prostatic carcinoma cell line orthotopically implanted in nude mice. Since *Hypericum perforatum* components act as serotonin-reuptake inhibitors and exert cytotoxic effects on several human cancer cell lines, in this work we analyzed the effect of a treatment with *Hypericum perforatum* extract on the growth of human prostate cancer cells *in vitro* and *in vivo.* This study highlighted a significant reduction of tumor growth and number of metastasis suggesting that this natural compound may be useful in the treatment of prostate cancer.[5]

References

1. Couldwell, W.T. *et al.* (1994). Hypericin: a potential antiglioma therapy. *Neurosurgery* 35: 705-710.

2. Vandenbogaerde, A de. and Witte, P. (1995). Antineoplastic properties of photosensitized hypericin (Meeting abstract). *In: Anticancer Res.* 15 (5A): 1757-8.

3. Alecu, M. *et al.* (1998). Photodynamic treatment of basal cell carcinoma and squamous cell carcinoma with hypericin. *Anticancer Research* 18: 4651-4654.

4. Schempp, C.M. *et al.* (2002). Inhibition of tumor cell growth by hyperforin, a novel anticancer drug from St. John's wort that acts by induction of apoptosis. *Onceogene* 21 (8): 1242-1250.

5. Martarelli, D., Martarellib, B., Pediconia, D., Nabissic, M.I., Perfumia M. and Pompeia. P. (2004). *Hypericum perforatum* methanolic extract inhibits growth of human prostatic carcinoma cell line orthotopically implanted in nude mice. *Cancer Letters* 210 (1): 27-33.

8

Hypericum and postmenopausal syndrome

Grube *et al.* (1999) studied the efficacy of *Hypericum perforatum* extract in the treatment of menopausal symptoms of psychological origin. One-hundred-and-six menopausal women between 43 and 65 years of age took a standardized extract of *Hypericum perforatum* for 12 weeks (one tablet, containing 300 mg of extract standardized to 0.3% hypericin, three times daily). The women were examined and returned questionnaires at baseline and at 5, 8 and 12 weeks. The incidence and severity of psychological (irritability, anxiety, etc.), psychosomatic (palpitations, sleep disturbances, headache) and vasomotor (hot flushes, dizziness) symptoms were recorded.

At baseline, the majority of symptoms were rated as 'marked' or 'moderate' by the participants. Improvements were seen already after 5 weeks, and after 12 weeks the majorities of women were symptom-free or rated the severity of their symptoms as 'slight'. The Menopausal Rating Scale, which rates the severity of 10 groups of menopausal symptoms, showed quite a dramatic and statistically significant fall over the 12-week period ($p<0.001$).

The study also assessed the effect of the extract of *Hypericum perforatum* on the participants' sexuality, using an 8-question questionnaire. Overall, about 80% of the women considered that the *Hypericum perforatum* treatment had substantially enhanced their sexuality. Both the examining physician and the women themselves also evaluated the therapeutic outcome of the treatment using the Clinical Global Impression Scale. Menopausal symptoms were found to have diminished or disappeared altogether in the majority of women (79.2% by physician's rating [p<0.001] and 76.4% by self-rating [p<0.001]).

Reference

1. Grube, B. *et al.* (1999). St John's wort extract: efficacy for menopausal symptoms of psychological origin. *Advances in Therapy* 16: 177-186

9

Antiviral Studies of Hypericum and Hypericin

Mishenkova *et al.* (1975) reported inhibited effect of flavonoids and catechins found in *Hypericum perforatum* on the influenza virus (83% to 100%) in laboratory experiments.[1]

Takahashi *et al.* (1989) investigated hypericin and pseudohypericin which have been isolated from plants of the Hypericum family are aromatic polycyclic diones. Daniel Meruelo *et al.* have reported that hypericin and pseudohypericin showed potent antiretroviral activity including anti-human immunodeficiency virus. However, the mechanism of these antiretroviral activities has not been clarified. In the course of screening specific inhibitors of protein kinase C we have found that both compounds specifically inhibit protein kinase C with IC50 values 1.7 micrograms/ml and 15 micrograms/ml, respectively, and show antiproliferative activity against mammalian cells. These data suggest that antiretroviral activity of hypericin and pseudohypericin could be attributable to the inhibition of some phosphorylation involved by protein kinase C during viral infection of cells.[2]

Mcauliffe *et al.* (1993) described a phase I study of intravenous dosed synthetic hypericin. HIV- infected patients with less than or equal to 300 CD4 cells were eligible for enrollment. Patients received intravenous hypericin at 0.25 or 0.5 mg/kg BIW or 0.25 mg/kg thrice a day; with assessments of toxicity, pharmacokinetic and antiviral activity. 4 patients received a single oral dose. 25 patients received hypericin from 1 to 24 weeks.

Phototoxicity developed in all patients, varied in severity, and was dose limiting at 0.5 mg/ kg twice a day. Pharmacokinetic showed area under the curve of 26.0 ± 5.0 (Mean ± SD) and 53.7 ± 27.2 ± micrograms/hr/ml at intravenous doses of 0.25 and 0.5 mg/kg, respectively; peak levels were 4.2 ± 1.1 and 7.7 ± 2.0 micrograms/ ml and fell below 1 microgram/ml at 5 and 11 hours. Elimination half lives were 23.7 ± 7.3 and 35.3 ± 9 hrs (p greater than 0.1). Oral bioavailability was 22.3 ± 7.7%. A consistent change in antiviral endpoints was not seen with intermittent IV dosing. Intravenous dosing of hypericin yields dose-limiting cutaneous phototoxicity of variable severity. Pharmacokinetic data predict chronic oral dosing should achieve sustained blood levels in an anti-retroviral range.[3]

Klose and Wernet (1993) in an open pilot study investigated efficacy of with Hyperforat i.v. 2x2 mL weekly plus 3x2 Hypericum tablets per day in 18 HIV patients. The 16/18 patients with good study compliance in their majority showed stable or even increasing counts of absolute CD4 values for helper T cells over the 40 months of observation until now. It is remarkable that this trend was apparently not dependent on the level of the absolute CD4 count at the beginning of the Hypericum treatment. Also the CD4/CD8 ratio showed an improvement in the majority of these patients. Clinically, it was noteworthy that only two of these 16 patients encountered an opportunistic infection during the 40 months of observation. The other 14/16 patients remained clinically stable and are active in work and life with a Karnovsky-Index of 100. This steady state situation of the HIV infection also correlated with stable values of hemoglobin, leukocytes and platelets.

Furthermore, none of the otherwise known viral complications due to CMV, herpes or EBV was encountered in these 16 patients. Also in no instance a toxoplasmosis nor neurological symptoms became apparent. Hypericum perforatum is presented as a novel effective anti-viral substance of broader activity, though with a hitherto unknown mode of action. It should be noted that no side effects have been seen or measured in any of these 16 patients until now. On this basis an extended trial with more patients in all CDC categories seems warranted now.[4]

Vonsover *et al.* (1996) investigated the antiretroviral efficacy of long term therapy with plant derived hypericin on p24 antigen and on viral load in AIDS patients. Eighteen seropositive symptomatic individuals received monotherapy with plant derived hypericin [intravenous 1x2 ml weekly plus 6x2 hypericum perforatum tablets per day] for periods ranging between 48-72 months. Plasma viral RNA loads were monitored using Nucleic Acid Amplification Assay (NASBA, Organon-Teknika), and p24 antigen with an HIV-1 Antigen Base Dissociation Assay (Organon-Teknika); they were followed during 40 months of therapy.

Among the 18 participants, 33% had measurable, pre-treatment concentrations of serum p24 antigen, one became positive after 24 months and twelve remained p24 antigen negative throughout the follow- up period. Four patients exhibited significant long term declines in p24 antigen levels; the levels remained unchanged in 2 others. HIV-1 RNA was detected in 83% of the patients with a mean of maximal RNA load of 5.33 and 5.00 log10 copies/ml in p24 antigen-positive or negative patients respectively.

A substantial decline in viral load (from 5.0 to 4.23 log 10 copies/ml) was observed in most of the p24 Ag. positive and negative individuals. In some patients an increase in virus load was observed throughout a three year treatment period without effect on the clinical conditions of viral CMV, herpes or EBV complications being encountered. The findings provide strong support to the use of hypericin or its plant-derived form (*Hypericum perforatum*) as non-toxic and effective long term antiviral treatment modalities. More extended trials with more patients of all CDC categories are in progress.[5]

Gulick *et al.* (1999) treated 30 HIV-positive patients with synthetic hypericin 0.25 or 0.5 mg/kg i.v. twice weekly or 0.25 mg/kg i.v. 3 μ /wk, or 0.5 mg/kg/day p.o. for 8 weeks). Owing to phototoxic effects, only 14 patients completed the study. No significant change was found in CD4 cell counts, HIV p24 antigen levels, or HIV RNA copies, and out of 23 patients available for evaluation, 11 developed severe cutaneous phototoxicity, including all who received oral hypericin.[6]

Koytchev (1999) studied the efficacy of *Hypericum perforatum* in a double-blind, placebo-controlled, randomized trial in 110 patients with herpes genitalis. Symptoms were significantly and equally reduced compared to placebo ($p<0.0001$), including severity of episodes, size of area affected, and numbers of blisters. The treatment period lasted 90 days with an oral dosage of 300 mg 3 times a day and 600 mg 3 times a day on the days of herpes episodes. In a parallel trial of the same design and duration in 94 patients with herpes labialis, the results were reported to be of the same significance with equal alleviation of the same symptoms.[7]

References

1. Mishenkova, E.L. *et al.* (1975). Antiviral properties of St. John's wort and preparations produced from it. *Tr S'ezda Mikrobiol Ukr.* 222-223.
2. Takahashi, I., Nakanishi, S., Kobayashi, E., Nakano, H., Suzuki, K. and Tamaoki, T. (1989). Hypericin and pseudohypericin specifically inhibit protein kinase C: possible relation to their antiretroviral activity. *Biochem Biophys Res Commun.* 165 (3): 1207-12.
3. Mcauliffe, V., *et al.* (1993). A phase I dose escalation study of synthetic hypericin in HIV infected patients (ACTG 150). *Natl Conf Hum Retroviruses Relat Infect* (1st) 12-16):159.
4. Klose, A. and Wernet, P. (1993). Successful long term treatment over 40 months of HIV-patients with intravenous Hypericin. *Int Conf AIDS.* 9 (1): 470 (abstract no. PO-B26-2012).
5. Vonsover, A. Steinbeck, KA. Rudich, C. Mazur, Y. Lavie, D. Mandel, M. and Lavie, G. (1996). HIV-1 virus load in the serum of AIDS patients undergoing long term therapy with hypericin. *Int Conf AIDS.* 11 (1): 120 (abstract no. MO.B.1377).
6. Gulick, R.M. *et al.* (1999). Phase I studies of hypericin, the active compound in St. John's wort, as an antiretroviral agent in HIV-infected adults. *Annals of Internal Medicine* 130: 510-514.
7. Koytchev, R., Alken, R.G. and Dundarov, S. (1999). Hypericum-extract LI 160 for the therapy of Herpes simplex genitalis and labialis: results of two placebo-controlled, randomized, double-blind clinical trials. *Zeitschrift für Phytotherapie* 20:92 (abstract in English).

10

Hypericum and Immunity

Evstifeeva and Sibiriak (1996) carried out a systematic study of immunotropic activity of John's wort on the level of integrated fractions including all basic active substances of this plant. Both types of substances capable of increasing and suppressing the immunity were found in John's wort. Polyphenol fraction exhibits the immunostimulating activity with respect to the system of mononuclear phagocyte system, cellular and humoral immunity and is capable of recovering the immune response in conditions of high-zone tolerance. Lipophilic fraction of this plant exhibits immunosuppressant properties with respect to cellular and humoral immune response. Immunotropic activity of polyphenol and lipophilic fractions correlate with their low toxicity.[1]

Reference

1. Evstifeeva, T.A. and Sibiriak, S.V. (1996). The immunotropic properties of biologically active products obtained from Klamath weed (*Hypericum perforatum* L.) *Eksp Klin Farmakol* 59(1): 51-4.

11

Hypericum and Inflammation

Shipochliev *et al.* (1981) reported anti-inflammatory activity of extract of *Hypericum perfortatum*. The extract suppressed leukocyte infiltration in mice treated with carageenan and PGE-1.[1] Ozturk *et al.* (1996) reported analgesic activity of *Hypericum perforatum*.[2]

Panossian *et al.* (1996) examined the effects of hypericin on the release of inflammatory mediators by human immune system cells. At low concentrations (IC50 4-8 M), hypericin significantly and dose dependently inhibited the release of arachidonic acid (AA) and, as a result, of leukotriene B4 (LTB4). The authors found that hypericin strongly inhibited interleukin-1 alpha (IL-1α), possibly by way of inhibiting protein kinase C (PKC), as shown by Takahashi *et al.* (1987) at about the same concentration (0.1-10 μM). IL-1 is well-known as a potent factor in mediating inflammation and fever, inducing acute-phase reaction, nitric oxide production, reactive oxygen intermediate leading to free radical production, and activates B cells, eosinophils, natural killer (NK) cells, granulocytes and T cells. However, an extract of St. John's wort proved far stronger than hypericin at inhibiting the release of IL-1 and of LTB4. The authors suggested that the inhibitions of these mediators' results in a quelling of the inflammatory response (immunosuppression), and that apparently besides hypericin, other active constituents are involved.[3]

Bork *et al.* (1999) investigated the effect of *Hypericum perfortatum* on the NF-μB inflammation factor was conducted by Bork *et al.*, 1999). Their *in vitro* experiments demonstrated that

that whereas a St. John's wort ethanolic extract standardized to contain hyperforin and hypericin (5% and 0.15%, respectively) showed no activity at 200 μg/mL, hypericin at 2 μg/mL proved a potent inhibitor of TNFμ-induced activation of NF-μB, and at a concentration comparable to that of a potent NF-μB inhibitor (parthenolide 5 μM) found in feverfew (*Tanacetum parthenium* [L.] Schulz Bip). Hypericin (2 μg/mL) also inhibited the transcription of IL-6 (an inflammatory cytokine), almost totally preventing the TNF-μ-induced expression of interleukin-6 (IL-6). Hyperforin, however, showed no appreciable activity, even at 20 μg/mL (37.3 μM). The researchers therefore concluded that anti-inflammatory effects of St. John's wort extract preparations containing hypericin appear to be at least partly due to the inhibition of NF-μB.[4]

References

1. Shiplochliev, T. *et al.* (1981). Anti-inflammatory action of a group of plant extracts. *Vet Med Nauki.* 18: 87-94.
2. Ozturk, Y. *et al.* (1996). Effects of *Hypericum perforatum* L. and *Hypericum calycinum* L. extracts on the central nervous system in mice. *Phytomedicine.* 3: 139-146.
3. Panossian, A.G. *et al.* (1996). Immunosuppressive effects of hypericin on stimulated human leukocytes: inhibition of the arachidonic acid release, leukotriene B4 and interleukin-1 (production, and activation of nitric oxide formation). *Phytomedicine.* 3: 19-28.
4. Bork, P.M. *et al.* (1999). Hypericin as a non-antioxidant inhibitor of NF-μB. *Planta Medica* 65: 297-300.

12

Pharmacokinetics of Hypericin and Hyperforin

Stock and Holzl (1991) tested a radioactive isotope of hypericin in mice and humans. The maximum blood level was at 6 hours. The blood level at 3.5 hours was 0.2% of the dose and at 8 hours was 1.9% of the initial dose.[1]

Kerb, Brockmoller, Staffeldt, Ploch and Roots (1996) investigated single-dose and steady-state pharmacokinetics of hypericin and pseudohypericin in 13 healthy volunteers by administration of *Hypericum perforatum* Oral administration of 250, 750, and 1,500 micrograms of hypericin and 526, 1,578, and 3,156 micrograms of pseudohypericin resulted in median peak levels in plasma (Cmax) of 1.3, 7.2, and 16.6 micrograms/liter for H and 3.4, 12.1, and 29.7 micrograms/liter for pseudohypericin, respectively. The Cmax and the area under the curve values for the lowest dose were disproportionally lower than those for the higher doses. A lag time of 1.9 h for H was remarkably longer than the 0.4-h lag time for pseudohypericin.

Median half-lives for absorption, distribution, and elimination were 0.6, 6.0, and 43.1 h after 750 micrograms of hypericin and 1.3, 1.4, and 24.8 h after 1,578 micrograms of pseudohypericin, respectively. Fourteen-day treatment with 250 micrograms of hypericin and 526 micrograms of pseudohypericin three times a day resulted in median steady-state trough levels of 7.9 micrograms/liter for hypericin and 4.8 micrograms/liter for pseudohypericin after7 and 4 days, respectively; the corresponding Cssmax levels were 8.8 and 8.5 micrograms/liter, respectively.

Kinetic parameters after intravenous administration of Hypericum extract (115 and 38 micrograms for hypericin and pseudohypericin, respectively) in two subjects corresponded to those estimated after an oral dosage. Both hypericin and pseudohypericin were initially distributed into a central volume of 4.2 and 5.0 liter, respectively. The mean distribution volumes at steady state were 19.7 liters for hypericin and 39.3 liters for pseudohypericin, and the mean total clearance rates were 9.2 ml/min for hypericin and 43.3 ml/min for pseudohypericin. The systemic availability of hypericin and pseudohypericin from LI 160 was roughly estimated to be 14 and 21%, respectively. Treatment with extract of *Hypericum perforatum*, even in high doses, was well tolerated.[2]

Biber, Fischer, Romer and Chatterjee, (1997) investigated the pharmacokinetics of hyperforin in healthy volunteers. After administration of a 300mg tablet of extract of *Hypericum perforatum* containing 14.8 mg hyperforin, a maximum plasma level of approximately 150 mg/mL (280nM) hyperforin was reached after 3.5 hours. The oral bioavailability of hyperforin in doses up to 30 mg (i.e. 600 mg St. John's wort extract) was high. The half-life of hyperforin was 9 hours and the mean residence time 12 hours. No accumulation of hyperforin occurred with repeated dosing. Estimated steady state plasma concentrations with 3x300 mg extract per day were approximately 100 ng/mL or 180nM. These data show that the oral bioavailability of hyperforin is high and that steady-state plasma concentrations can easily be achieved and maintained with a three times daily dosing schedule.[3]

Cui *et al.* (2004) investigated in vitro metabolism profile of hyperforin using liver microsomes from male and female Sprague-Dawley rats, with or without induction by phenobarbital or dexamethasone. Four major Phase I metabolites, named 19-hydroxyhyperforin, 24-hydroxyhyperforin, 29-hydroxyhyperforin, and 34-hydroxyhyperforin, were isolated by high performance liquid chromatography and identified by mass spectrometry and NMR. Results suggest that hydroxylation is a major biotransformation of the hyperforin pathway in rat liver and that inducible cytochrome

P450 3A (CYP450 3A) and/or CYP2B may be the major cytochrome P450 isoforms catalyzing these hydroxylation reactions.[4]

References

1. Stock, S. and Holz, J. (1991). Pharmacokinetic test of (14 C)-labeled hypericin and pseudohypericin from Hypericum perforatum and serum kinetics of hypericin in man. *Planta Med* 57(2): 61-62.
2. Kerb, R., Brockmoller, J., Staffeldt, B., Ploch, M. and Roots, I. (1996). Single-dose and steady-state pharmacokinetics of hypericin and pseudohypericin. *Antimicrob Agents Chemother* 40(9): 2087-93.
3. Biber, A., Fischer, H., Romer, A. and Chatterjee, S.S. (1998). Oral bioavailability of hyperforin from hypericum extracts in rats and human volunteers. *Pharmacopsychiatry*. 31: 36-43.
4. Cui, Y., Ang, C.Y., Beger, R.D., Heinze, T.M., Hu, L. and Leakey, J. (2004). In vitro metabolism of hyperforin in rat liver microsomal systems. *Drug Metab Dispos* 32(1): 28-34.

13

Drug Interactions of *Hypericum perforatum*

Hoelzl & Ostrowski, (1986) are of the view that simultaneous use of *Hypericum perforatum* and a monoamine oxidase inhibitor may lead to increased effects and possible toxicity in form of high blood pressure. *Hypericum perforatum* has slight monoamine oxidase inhibiting activity. This information is based on animal studies, however clinical data is missing.[1]

Schaefer, (1996) is of the view that simultaneous use of *Hypericum perforatum* and a selective serotonin reuptake inhibitor may lead to an increased effect and possible toxicity (serotonin syndrome). *Hypericum perforatum* has slight serotonin reuptake properties. This information like the first is based on animal studies and no clinical study has been done.

Barone *et al.* (2000) reported that a 29-year-old white woman who received a cadaveric kidney and pancreas transplant, with stable organ function and stable cyclosporine concentrations began self-medicating with *Hypericum perforatum*. After taking *Hypericum perforatum* supplements for four to eight weeks, her cyclosporine concentrations became sub therapeutic; this was associated with organ rejection.

Four weeks after stopping *Hypericum perforatum*, her cyclosporine concentrations again became therapeutic. Subsequent to this rejection episode, she has developed chronic rejection and now has returned to dialysis. *Hypericum perforatum* is suspected to be a significant inducer of CYP3A4 isoenzyme activity and of P-glycoprotein (P-gp) expression, both of which are important in the metabolism and absorption of cyclosporine.

Cyclosporine exhibits a relatively small therapeutic window and is sensitive to medications that can modulate the CYP3A4 isoenzyme and P-gp in both the liver and small intestines. Patients taking *Hypericum perforatum* concomitant with other prescription medications whose absorption and metabolism are mediated by the CYP3A4 isoenzyme and P-gp require close monitoring.[2] The Food and Drug Administration issued a public health advisory in February 2000 concerning the probable interaction between indinavir and *Hypericum perforatum* that resulted in significantly decreased blood levels of this protease inhibitor, a class of medications used to treat HIV or AIDS. The Food and Drug Administration recommend that *Hypericum perforatum* not be used with any type of antiretroviral medication used to treat HIV or AIDS.[3]

Hypericum perforatum should not be taken by those on digoxin because the herb may decrease levels of the medication and reduce its effectiveness.[4] Potential drug interaction has been reported between *Hypericum perforatum* and theophylline.[5] Burstein *et al.* (2000) reported that *Hypericum perforatum* has no effect on pharmacokinetic profile of carbamezapine in healthy volunteers.[6]

References

1. Hoelzl, J., Demisch, L. and Gottnik, B. (1989). Investigations about antidepressive and mood changing effects of *Hypericum perforatum*. *Planta Med.* 55: 643.

2. Barone, G.W., *et al.* (2000). Drug interaction between St. John's wort and cyclosporine. *Ann Pharmacother* 34: 1013-1016.

3. Food and Drug Administration. *Risk of Drug Interactions with St John's Wort and Indinavir and Other Drugs*. Rockville, Md: National Press Office; Public Health Advisory.

4. Johne, A., *et al.* (1999). Pharmacokinetic interaction of digoxin with an herbal extract from St. John's wort (*Hypericum perforatum*). *Clin Pharmacol Ther* 66: 338-345.

5. Nebel, A., *et al.* (1999). Potential metabolic interaction between St. John's wort and theophylline. *Ann Pharmacother* 33: 502.

6. Burstein, A.H., *et al.* (2000). Lack of effect of St. John's wort on carbamezapine pharmacokinetics in healthy volunteers. *Clinical Pharmacology and Therapeutics* 68: 66.

14

Adverse Drug Reactions and Toxicity of Hypericum

Pace and MacKinney (1941) showed that Hypericin in concentrations as low as 10(-7), can lysis red blood cells with exposure to light of the correct wavelength.[1] Schiplochliev, (1981) states that Hypericum has slight uterotonic actions *in vitro*.[2]

Garrett *et al.* (1982) reported the effect of *Senecio jacobaea, Symphytum officinale, Hypercium perforatum, Pteridium aquilinum* and *Medicago sativa* on hepatic drug-metabolizing enzymes in rats were measured. *Senecio jacobaea* and *Pteridium aquilinum* increased (P less than 0.05) the activity of glutathione transferase and epoxide hydrolase. *Symphytum officinale* and *Medicago sativa* increased (P less than 0.05) the activity of aminopyrine N-demethylase. *Pteridium aquilinum* or *Hypercium perforatum* in conjunction with *Senecio jacobaea*, did not influence chronic toxicity of *Senecio jacobaea*, as assessed by rat survival time. *Senecio jacobaea*, resulted in increased (P less than 0.05) hepatic copper levels; the other plants did not affect copper levels. The results do not suggest any major interaction in the toxicity of *Senecio jacobaea, Hypercium perforatum* or *Pteridium aquilinum*.[3]

Duran & Song (1986) reported photosensitivity with sunburn-like lesions and inflammation of the mucous membranes, at least in animals.[4] Kumper (1989) reported about a case of *Hypercium perforatum* poisoning in German Blackface sheep is given. After the ingestion of *Hypericum perforatum* all slightly pigmented parts of the skin that were rarely covered with hair were photosensitized. In summer many sheep suffered from inflammatory skin alterations at the ears, the bridge of the nose and at the surroundings of the eyes.[5]

Southwell & Campbell (1991) described hypericism, which is sensitivity to light observed in animals that ingest certain species of Hypericum. The pigments are carried to the skin, and in the unpigmented, unhaired portions of the skin of sheep, cattle, horses, goats and swine cause a central nervous system depression, an increased sensitivity to stress (like cold, heat, handling), and produced sun-light induced rash and blisters.[6]

Acute toxicity studies using rats, guinea pigs, and mice indicate that the extract is relatively nontoxic. Animals fed with *Hypercium perforatum* flowers for 2 weeks showed significant signs of toxicity, including erythema, edema of the portion of the body exposed to light, alopecia, and changes in blood chemistry. In a chronic study, rats fed with *H. perforatum* gained less weight than control animals. Mixtures containing the extract and the oil were not irritants or sensitizers in animals.

Mixtures of the extract and the oil produced minimal or no ocular irritation in rabbit eyes. Mutagenic activity in an Ames test was attributed to flavonols in one study and to quercitin in another, but other genotoxicity assays were negative.

No carcinogenicity or reproductive and developmental toxicity data were available. A mixture of the extract and the oil was not irritating in clinical studies. Adverse reactions to Hypericum extract in the clinical treatment of depression include skin reddening and itching, dizziness, constipation, fatigue, anxiety, and tiredness.[7]

Kako and Saleem (1993) reported a study of sheep's experimentally poisoned with *Hypericum perforatum*. Sheep given different dosages and frequencies of *Hypericum perforatum* had decreased hemoglobin, red blood cell count, packed cell volumes, total protein, glucose, cholesterol, triglycerides, and serum alkaline phosphotase activities. Blood urea nitrogen, sodium, potassium, bilrubin (total and direct), and the activities of aspartate aminotransferase, alanine aminotransferase, lactate dehydrogenase and gamma glutamyltransferase increased.[8]

Woelk *et al.* (1994) reported restlessness (0.3%) and fatigue (0.4%) in 3,250 patients in one study of depressed patients[9]. Roots *et al.* (1996) reported photosensitization in a controlled clinical trial using metered doses of hypericin and subsequent exposure to UVA/UVB radiation.[10] Vorbach *et al.* (1997) recorded dry mouth and constipation as side effects associated with Hypericum therapy.[11] Wheatley (1998) reported headache, anorexia, diarrhoea, nausea and stomachache as side effects associated with *Hypericum perforatum.*[12]

Leuschner (1997) reported when 300, 900, and 2700 milligrams/kilogram (mg/kg) were given to rats and dogs for 26 weeks, the lowest dose which showed any toxicity was 900 mg/kg, which produced a non-specific syndrome of reduced weight, changes in the clinical chemical parameters, slight kidney damage, slight liver damage and blood changes. The animals were examined histologically and found to have mild hypertrophy of zona glomerulosa and adrenals. All changes returned to normal if the treatment was stopped for 4 weeks.[13] It suggests that flavonoids such as quercetin (constituent of *Hypericum perforatum* extract), have genotoxic potential. Oil of Hypericum has been found to be slightly mutagenic.[14]

References

1. Pace, N. and MacKinney, G. (1941). Hypericin, the photodynamic pigment from St. John's Wort. *J Am Chemical Soc.* 63: 2570-2574.
2. Shiplochliev, T. (1981). Extracts from a group of medicinal plants enhancing the uterine tonus. *Vet Med Nauki* 1981; 18: 94-98.
3. Garrett, B.J., Cheeke, P.R., Miranda, CL., Goeger, D.E. and Buhler D.R. (1982). Consumption of poisonous plants (*Senecio jacobaea, Symphytum officinale, Pteridium aquilinum, Hypericum perforatum*) by rats: chronic toxicity, mineral metabolism, and hepatic drug-metabolizing enzymes. *Toxicol Lett* 10 (2-3): 183-8.
4. Duran, N. and Song, P.S. (1986). Hypericin and its photodynamic action. *Photochem Photobiol.* 43: 677-680.
5. Kumper, H. (1989). Hypericum poisoning in sheep. *Tierarztl Prax* 17 (3): 257-61.
6. Southwell, I.A. and Campbell, M.H. (1991). Hypericin content variation in *Hypericum perforatum* in Australia. *Phytochem* 30:475-478.
7. No authors listed (2001). Final report on the safety assessment of *Hypericum perforatum* extract and *Hypericum perforatum* oil. *Int J Toxicol* 20 (2): 31-39.

8. Kako, M.D., al-Sultan and Saleem, A.N. (1993). (Studies of sheep experimentally poisoned with *Hypericum perforatum. Vet Hum Toxicol* 35(4): 298-300.

9. Woelk, H. Burkard, G. and Gruenwald, J. (1994). Benefits and risks of the *Hypericum* extract LI 160: drug monitoring with 3250 patients. *J Geriatric Psychiatry Neurol.* 7 (1): 34-38.

10. Roots, *et al.* (1996). Photosensitization has been demonstrated in a controlled clinical trial using metered doses of hypericin and subsequent exposure to UVA/UVB radiation.

11. Vorbach, E.U., Huebner, W.D. and Arnoldt, K.H. (1994). Effectiveness and tolerance of the hypericum extract LI 160 in comparison with imipramine: randomized double-blind double-blind study to 135 out patients. *J Geriatr Psychiatry Neurol* 7 (1): 19-23.

12. Wheatley, D. (1997). LI 160, and extract of St. John's Wort, versus amitriptyline in mildly to moderately depressed outpatients - a controlled 6-week clinical trial. *Pharmacopsychiatry* 30 (2): 77-80.

13. Leuschner, G. (1996/1997). Preclinical toxicology profile of *Hypericum extracts* LI 160 (abstract). Phytomedicine *Int J Phytother Phytopharmacol* 3(1): 104.

14. Poginsky, B., *et al.* (1988). Johanniskraut (*Hypericum perforatum* L.) Genotoxizitaet bedingt durch den Quercetingehalt. *Dtsch Apoth Ztg* 128: 1364-1366.

15

Investigations of Other Hypericum species

Hypericum chinese Linn.

Tada, Chiba, Takakuwa and Kojima, (1992) studied the analogues of natural phloroglucinols as antagonists against both thromboxane A2 and leukotriene D4. Antagonists were prepared from phloroglucinol. The correlation between the structures and activity was studied in the synthesized and naturally occurring phloroglucinol derivatives.[1]

Hypericum drummondii (Grev. & Hook.) T. & G.

1. Jayasuriya, Clark and McChesney (1991) reported that bioactivity-guided fractionation of the hexane extract of the stems and leaves of *Hypericum drummondii* afforded four new filicinic acid derivatives: drummondin D, isodrummondin D, drummondin E and drummondin F. All compounds possessed strong antibiotic activity against the Gram-positive bacteria *Staphylococcus aureus* and *Bacillus subtilis* and the acid fast bacterium *Mycobacterium smegmatis.*[2]
2. Jayasuriya, Machesney, Swanson and Pezzuto (1989) reported significant activity of hexane extracts of *Hypericum drummondii* against the Gram-positive bacteria *Staphylococcus aureus, Bacillus subtilis*, and the acid-fast bacterium Mycobacterium smegmatis in an agar

well diffusion assay. Bioassay-directed fractionation afforded four new rottlerin-type compounds (drummondins A, B, C and F. The antimicrobial activity of these compounds was comparable to or greater than that demonstrated by streptomycin and generally correlated with cytotoxic activity determined with cultured P-388, KB, or human cancer cell lines (breast, colon, lung, melanoma). No cell-type selectivity was observed. In addition, two known compounds, albaspidins A-A and P-P, were isolated and structured characterized. Neither demonstrated appreciable antimicrobial or cytotoxic activity.[3]

3. Jayasuriya, Baker, Clark and McChesney (1991) reported Drummondin C, an antibiotic isolated from a bioassay-directed fractionation of *Hypericum drummondii.* It showed significant activity against the Gram-positive bacteria *Staphylococcus aureus* and *Bacillus subtilis* and the acid-fast bacterium *Mycobacterium smegmatis.* Two structural units of drummondin C, the 8-acetyl-5, 7-dihydroxy-2, 2-dimethylchromene and 5-acetyl-3- methyl-filicinic acid, were synthesized to determine the relative importance of the two substructure portions to the antibiotic activity of the compound. The low antimicrobial activity of 6 and 9 demonstrates the necessity of both units for the antibiotic activity of drummondin C.[4]

Hypericum roeperianum Schimp. ex A. Rich

Rath, Potterat, Mavi and Hostettmann, (1996) reported about isolation of four new xanthones from the roots of *Hypericum roeperianum*. Some of the isolated xanthones exhibited antifungal activity against *Candida albicans.*[5]

Hypericum brasiliense Choisy

1. Rocha *et al.* (1994) isolated a new gamma-pyrone (hyperbrasilone), three known xanthones (1, 5-dihydroxyxanthone, 5-hydroxy-1-methoxyxanthone and 6-deoxyjacareubin) and betulinic acid from a dichloromethane extract of stems and roots of *Hypericum brasiliense.* Hyperbrasilone and the xanthones depicited antifungal activity against *Cladosporium cucumerinum,*

while the three xanthones showed differing degrees of inhibition of monoamine oxidase A and B.[6]

2. Rocha *et al.* (1995) isolated three known phloroglucinols (japonicine A, uliginosin A and isouliginosin B) and a new phloroglucinol (hyperbrasiol A) from a petrol extract of the leaves and flowers of *Hypericum brasiliense*. All four phloroglucinols exhibited antibacterial activity against *Bacillus subtilis*. The flavonoids, kaempferol, luteolin, quercetin, quercitrin, isoquercitrin, hyperoside and guaijaverin, were isolated from a methanol extract of the leaves and flowers.[7]

Hypericum calycinum DC and *Hypericum revolutum* Vahl

Decosterd *et al.*, 1988 reported *in vitro* cytotoxic actions of hyperforin isolated from *Hypericum calycium* and *Hypericum revolutum* against human colon carcinoma cells.[8]

Decosterd, Hoffmann, Kyburz, Bray and Hostettmann (1991) isolated a new phloroglucinol derivative from the light petroleum ether extract of the aerial parts of *Hypericum calycinum*. It has fungicidal activity against *Cladosporium cucumerinum* In addition; this new phloroglucinol derivative has antimalarial activity in an (*in vitro*) test system.[9]

Hypericum japonicum Thunb

Ishiguro *et al.* (1990) isolated a new antibiotic compound, sarothralin G from *Hypericum japonicum* Thunb. The compound contains phloroglucinol and filicinic acid moieties.[10]

Hypericum patulum Thunb and *Hypericum mysorense*

1. Ishiguro *et al.* (1999) isolated two new xanthone glycosides, patuloside A and patuloside B from cell suspension cultures of *Hypericum patulum*.[11]
2. Mukherjee *et al.* (2001) investigated the wound healing potential of methanol extract of leaves of *Hypericum patulum* in different experimental models of wounds in rats. The methanol extract of leaves (HPM), in the form of an ointment with two different concentrations (5% and 10% w/w ointment of leaf extract in simple ointment base)

was evaluated for wound healing potential in an excision wound model and an incision wound model in rats. Both concentrations of the methanol extract ointment showed significant responses in both the wound types tested when compared with the control group. The effect produced by the extract ointment, in terms of wound contracting ability, wound closure time, regeneration of tissues at wound site, tensile strength of the wound and histopathological characteristics were comparable to those of a standard drug nitrofurazone ointment.[12]

3. Mukherjee *et al.* (2001) investigated the antimicrobial activity of petroleum ether, acetone, chloroform and methanol extracts of the leaves and stems of *Hypericum mysorense* and *Hypericum patulum* by the disc diffusion method. The results showed that both the stem and leaf extracts of both species have a very broad spectrum of antibacterial activity. The petroleum ether and acetone extract of the leaves and stems of *Hypericum mysorense* did not show antifungal effects against any of the six fungal organisms tested but other extracts showed a potential antifungal effect which was comparable to that of griseofulvin, the standard antifungal agent. Thus these results demonstrate that the leaf and stem extracts of both the Hypericum species have a very broad spectrum of activity and suggest that they may be useful in the treatment of various microbial infections.[13]

Hypericum canariense Linn.

Rabanal *et al.* (2002) investigated the traditional use of *Hypericum canariense* for skin infections.The plant exhibited antibacterial activity against several test organisms. However no antifungal activity could be demonstrated with the plant.[14]

Hypericum erectum Thunb

Perry & Metzger (1980) repoterf that sap from the fresh leaves of *Hypericum erectum* is used on cuts and bruises. The decoction of the mature dried plant acts as a styptic on wounds. A poultice of the crushed leaves and green stem is applied to animal bites and insect stings.[15]

Hypericum hookerianum W & A

Mukherjee & Suresh (2000) investigated the traditional use of *Hypericum hookerianum* as a wound healing agent. Ointment prepraed form the leaf extract significantly enhanced various healing parameters in two experimental rat wound healing models.[16]

Hypericum androsaemum

Valentao *et al.* (2004) reported hepatotoxic effect of infusion of *Hypericum androsaemum*. However, the results showed that the oral pretreatment with *Hypericum androsaemum* infusion (4, 20 and 100 mg/kg) for 4 days before a single intraperitoneal dose of t-BHP (1.8 mmol/kg) potentiated the t-BHP-induced hepatotoxicity. In fact, it was observed a potentiation in the depletion of total glutathione and reduced glutathione (GSH) contents and increase in oxidised glutathione (GSSG) level. Also the histopathological evaluation of the mice livers revealed that the infusion raised the incidence of liver lesions induced by t-BHP. These data do not corroborate any effect of *Hypericum androsaemum* infusion as hepatoprotector, but rather as a potentiator of hepatotoxicity in the present experimental conditions.[17]

References

1. Tada, M., Chiba, K., Takakuwa, T. and Kojima, E. (1992). Analogues of natural phloroglucinols as antagonists against both thromboxane A2 and leukotriene D4. *J Med Chem* 35 (7): 1209-1212.
2. Jayasuriya, H., Clark, A.M. and McChesney, J.D. (1991). New antimicrobial filicinic acid derivatives from *Hypericum drummondii*. *J Nat Prod* 54 (5):1314-1320.
3. Jayasuriya, H., McChesney, J.D., Swanson, S.M. and Pezzuto J.M. (1989). Antimicrobial and cytotoxic activity of rottlerin-type compounds from *Hypericum drummondii*. *J Nat Prod* 52 (2): 325-331.
4. Jayasuriya, H., Baker, J.K., Clark, A.M., and McChesney J.D. (1991). Synthesis and the biological evaluation of the structural units of drummondin C. *Pharm Res* 8 (11): 1372-1376.
5. Rath, G., Potterat, O., Mavi, S. and Hostettmann, K. (1996). Xanthones from *Hypericum roeperanum*. *Phytochem* 43 (2): 513-520.

6. Rocha, L., Marston, A., Kaplan, M.A., Stoeckli-Evans, H., Thull, U., Testa, B. and Hostettmann, K. (1994). An antifungal gamma-pyrone and xanthones with monoamine oxidase inhibitory activity from *Hypericum brasiliense*. *Phytochem* 36 (6): 1381-1385.

7. Rocha, L., Marston, A., Potterat, O., Kaplan, M.A., Stoeckli-Evans, H. and Hostettmann, K. (1995). Antibacterial phloroglucinols and flavonoids from *Hypericum brasiliense*. *Phytochem* 40 (5): 1447-1452.

8. Decosterd, L.A., *et al.* (1988). Isolation of new cytotoxic constituents from *Hypericum revolutum* and *Hypericum calycinum* by liquid-liquid chromatography. *Planta Med* 54: 560.

9. Decosterd, L.A., Hoffmann, E., Kyburz, R., Bray, D. and Hostettmann, K. (1991). A new phloroglucinol derivative from *Hypericum calycinum* with antifungal and in vitro antimalarial activity. *Planta Med* 57 (6): 548-551.

10. Ishiguro, K., Yamaki, M., Kashihara, M., Takagi, S., Isoi, K., and Sarothralin, G. (1990). A new antimicrobial compound from *Hypericum japonicum*. *Planta Med* 56 (3): 274-276.

11. Ishiguro, K., Yamamoto, R. and Oku, H. (1999). Patulosides A and B, novel xanthone glycosides from cell suspension cultures of *Hypericum patulum*. *J Nat Prod* 62 (6): 906-908.

12. Mukherjee, P.K., Verpoorte, R. and Suresh, B. (2000). Evaluation of *in-vivo* wound healing activity of *Hypericum patulum* (Family: hypericaceae) leaf extract on different wound model in rats. *J Ethnopharmacol*70 (3): 315-321.

13. Mukherjee, P. K., Saritha, G. S. and Suresh, B. (2002). Antimicrobial potential of two different Hypericum species available in India. *Phytother Res* 16(7): 692-695.

14. Rabanal, R.M., Arias, A., Prado, B., Hernandez-Perez, M. and Sanchez-Mateo, C.C. (2002). Antimicrobial studies on three species of *Hypericum* from the Canary Islands. *J Ethnopharmacol* 81(2): 287-292.

15. Perry, L.M. and Metzger, J. (1980). Medicinal Plants of East and Southeast Asia: Attributed Properties and Uses. Cambridge, Massachusetts: MIT Press.

16. Mukherjee, P.K. and Suresh, B. (2000). The evaluation of wound-healing potential of *Hypericum hookerianum* leaf and stem extracts. *J Alter Compl Med* 6 (1): 61-69.

17. Valentao, P., Carvalho, M., Carvalho, F., Fernandes, E., das Neves R.P., Pereira, M.L., Andrade, P.B., Seabra, R.M. and Bastos, M.L. (2004). *Hypericum androsaemum* infusion increases tert-butyl hydroperoxide-induced mice hepatotoxicity *in vivo*. *J Ethnopharmacol* 94 (2-3): 345-351.

16

Oil of Hypericum

Oil of hypericum is of brilliant red colour. In 1616, hypericum oil was listed in the first London Pharmacopoeia as "Oleum Hyrerici," and was said to have great wound healing capabilities. Miller (1924) reported that oil prepared by distillation (method unknown) and let stand for 18 months in a full, well-stoppered glass bottle. It was kept in a cool, dark place. At the end of this time no change was noted in appearance or color, but analysis of the constituents showed changes in composition.[1] Wren (1975) says that the fresh flowers of *Hypericum perforatum* in olive oil make a healing application to wounds, sores, ulcers, and swellings.[2]

Araya and Ford (1981) in one extensive study, up to 80% of the hypericin was destroyed by drying of the fresh plant in sunlight. For this reason, modern herbalists generally grind the fresh tops of Hypericum and immediately macerate them in olive oil or sunflower seed oil. The oil is then pressed and filtered after two weeks, and should be stored in amber bottles away from heat and light.[3]

Maisenbacher and Kovar (1992) in a study showed that a sample containing 62 milligrams of hyperforin (the active ingredient in the oil) contained no hypericin in 14 days. If sunlight was not used to prepare the oil the breakdown was slower, but still less than 30 days. Various oil preparation methods have been described, including one with eutanol G which showed stability for 6 months. These authors evaluated 6 commercial samples of oil of

Hypercium containing 2.2 to 20.8 milligrams/deciliter. All hyperforin was gone by the end of five weeks.[4]

Gudzic *et al.* (2001) investigated the essential oil of *Hypericum olympicum* L. for the first time and was compared it with essential oil of *Hypericum perforatum* L. from the same location. Eighteen components were common for both species. The main components of *H. olympicim* oil were (*E*)-anethole (30.7%) and β-farnesene (12.4%), and for *H. perforatum* oil they were β-caryophyllene (14.2%) and 2-methyl-octane (13.1%).[5]

Goskel *et al.* (2004) studied antimicrobial activity of essential oils of *Hypericum scabrum*, *Hypericum scabroides* and *Hypericum triquetrifolium* against nine organisms. All the essential oils exhibited some broad spectrum antibacterial activity, at a concentration of 80 μg/mL. The essential oils of Hypericum species showed antibacterial activity against the tested organisms and a yeast.[6]

References

Miller, E.R. (1924).The volatile oil of *Hypericum perforatum*. *J Amer Pharm Assoc* 16: 824-828.

Wren, R.C. (1975). Potter's *New Cyclopaedia of Botanical Drugs and Preparations*. Bradford, Devon: Health Science Press.

Araya, O.S. & E.J.H. Ford (1981). "An investigation of the type of photosensitization caused by the ingestion of St. John's wort (*Hypericum perforatum*) by calves," *J Comps Path* 91: 135-41.

Maisenbacher, P. & Kovar, K.A. (1992). Analysis and stability of Hyperici Oleum. *Planta Med.* 58: 351-354.

Gudzic, B., Dordevic, S., Palic, R. and Stagampvoic (2001). Essential oils of *Hypericum olympicum* L. and *Hypericum perforatum* L. *Flavour and Fragrance Journal*. 16 (3): 201 – 203.

Göksel, K., Toker, Z., Ozen, H.C. and Aytekin, C. (2004). The antimicrobial activity of essential oils of *Hypericum scabrum, Hypericum scabroides* and *Hypericum triquetrifolium*. *Phytother Res* 18(4):339–341.

17

Dose of Hypericum

Extract: 300 milligrams of the standardized extract (0.3% hypericin content) taken three times a day. This is equivalent to 200 to 1000 micrograms/day of hypericin. This is applicable to tablet or capsule formation.[1]

Liquid extract: 2 to 4 milliliters, 3 times a day.[2]

Tincture: 2 to 4 milliliters, 3 times a day.[2]

References

1. Muller, J.L. and Clauson, K.A. (1997). Pharmaceutical considerations of common herbal medicine. *Am J Man Care* 3:1753-1170.
2. Schmidt, U. and Sommer, H. (1993). Johanniskraut-Extract zur ambulanten Therapie der Depression. *Fortschr Med* 111: 339-342

18

Emerging Concepts

Hypericum and Cytochrome P450

Moore *et al* (2000) has shown that hyperforin and St John's Wort extracts induce CYP3A4 in hepatocyte cells via the pregnane X nuclear receptor which controls CYP3A4 expression.[1]

Obach (2000) examined the potential of Hypericum extracts to inhibit human cytochrome P450 (CYP) enzyme activities. The systems included were CYP1A2, CYP2C9, CYP2CI9, CYP2D6 and CYP3A4. The crude extracts of Hypericum perforatum demonstrated inhibitory effect on all the five systems with more sensitive action on CYP2C9, CYP2CI9, CYP2D6 and CYP3A4. The extracts were further fractioned and each fraction was tested for inhibitory action on the cytochrome P450 (CYP) enzyme activities.

Hyperforin, amentoflavone, hypericin, quercetrin and chlorogenic acid, all demonstrated inhibitory affect on cytochrome P450 (CYP) enzyme activities. Amentoflavone was shown to be potent inhibitor of CYP3A4, CYP1A2 and CYP2C9. Hyperforin was shown as potent non-competitive inhibitor of CYP2C9, CYP2D6 and CYP3A4. Hypericin also demonstrated inhibitory activity. The author concluded that *Hypericum perforatum* extracts contains compound that have potent inhibitory activity on major drug metabolizing enzymes. Further stress has been laid on pharmacokinetic studies of *Hypericum perforatum* so that the topic of drug interaction is not overlooked.[2]

Markowitz *et al.* (2000) investigated the effect of *Hypericum perforatum* on cytochrome CYP2D6 and CYP3A4. Seven healthy volunteers were orally administered alprazolam and dextromethorphan along with Hypericum perforatum. Urinary concentrations of alprazolam and dextromethorphan and dextromethorphan metabolic ratios were determined. The authors concluded no significant difference in pharmacokinetic profile of alprazolam or dextromethorphan metabolic ratios.[3]

Roby *et al.* (2000) in a human study was conducted on 13 healthy volunteers given 300mg standardized extract St John's Wort TID for 14 days. 24 hr urinary excretion ratios of 6-beta-hydroxycortisol/ cortisol were used as an index of CYP3A4 activity. A significant increase, from zero up to around 2.5x over base was found in urinary ratios in 12 subjects, suggestive of 3A4 induction.[4]

According to a study done by Wisley *et al.*, (Researchers of Glaxo-Smith line), they were able to determine structure of a significant molecule (PXR) present in the liver and responsible for metabolism of majority of the drugs consumed by the humans. The molecule is regulator of protein known as P450-3A, which is responsible for breaking the drugs. Recently interactions of *Hypericum perforatum* have been suggested with oral contraceptives, antivirals and transplant drugs.[5]

Cantoni *et al.* (2003), reported that hyperforin contributes to the hepatic CYP3A-inducing effect of *Hypericum perforatum* extract in the mouse. A hydroalcoholic extract containing 4.5% hyperforin was given at a dose of 300 mg/kg b.i.d. for 4 and 12 days. Hyperforin, its main phloroglucinol component, was given as dicyclohexyl ammonium salt (18.1 mg/kg, b.i.d.) on the basis of its content in the extract, to ensure comparable exposure to hyperforin. The extract increased hepatic erythromycin-N-demethylase activity, which is cytochrome P450 enzyme (CYP) 3A-dependent, about 2.2-fold after four days of dosing, with only slightly greater effect after 12 days (2.8 times controls). Hyperforin too increased erythromycin-N-demethylase activity activity within four days much the same extent as the extract (1.8 times the activity of controls), suggesting that it behaves qualitatively and quantitatively like the extract as regards induction of CYP3A activity. This effect was confirmed by Western blot analysis of hepatic CYP3A expression. Exposure to hyperforin at the end of the four days' treatment was still similar to that with extract of *Hypericum perforatum*, although it was variable and lower than after the first dose in both cases, further suggesting that hyperforin plays a key role in CYP3A induction by the extract of *Hypericum perforatum* in the mouse. Standardization of the extracts based on the hyperforin content can be proposed for further evaluation of their potential action on first-pass metabolism and clearance of co-administered CYP3A substrates.[6]

Watkins *et al.* (2003) in a study showed that hyperforin induces the expression of numerous drug metabolism and excretion genes in primary human hepatocytes. The researchers determined the crystal structure of hyperforin in complex with the ligand binding domain of human PXR. Hyperforin induces conformational changes in PXR's ligand binding pocket relative to structures of human PXR elucidated previously and increases the size of the pocket by 250 A(3). It was found that the mutation of individual aromatic residues within the ligand binding cavity changes PXR's response to particular ligands. Taken together, these results demonstrate that PXR employs structural flexibility to expand the chemical space it samples and that the mutation of specific residues within the ligand binding pocket of PXR tunes the receptor's response to ligands.[7]

Recently it has been reported that *Hypericum perforatum* can reduce the efficacy of chemotherapy drug known as irinotecan. The research was presented at the 93rd Annual Meeting of the American Association for Cancer Research by Ron A.H.J. Mathijssen, M.D., of the Rotterdam Cancer Institute in the Netherlands. Mathijssen's study was based on previous findings that *Hypericum perforatum* can stimulate cytochrome P450 (CYP3A4), an enzyme system involved in drug metabolism. Specifically, researchers went on the premise that hypericin and hyperforin-components of *Hypericum perforatum* -seem to enhance CYP3A4 activity, and the fact that irinotecan is also partly metabolized by the enzyme system.

Note: For more information on drug interactions, please refer to chapter 12.

References

1. Moore, L.B., *et al.* (2000) St. John's wort induces hepatic drug metabolism through activation of the pregnane X receptor. *Proc Natl Acad Sci* 97(13): 7500-7502.
2. Obach, R.S. (2000). Inhibition of human cytochrome P450 (CYP) enzymes by constituents of St. John's Wort, an herbal preparation used in the treatment of depression. *J Pharmacol Exp Ther.* 294(1): 88-95.
3. Markowitz, J.S. *et al.* (2000). Effect of St.John's wort on cytochrome P-450 2D6 and 3A4 activity in healthy volunteers. *Life Sciences* 66(9): 133-9.
4. Roby, C.A., *et al.* (2000). St John's Wort: effect on CYP3A4 activity. *Clin Pharmacol Ther.* 67(5): 451-7.
5. Williamson, D. (2001). Study uncovers structure of key molecule responsible for clearing drugs form body. *Carolina News service.* 294.
6. Cantoni, L., Rozio, M., Mangolini, A., Hauri, L. and Caccia, S. (2003). Hyperforin contributes to the Hepatic CYP3A-inducing effect of *Hypericum perforatum* extract in the Mouse. *Biotransformation and Toxicokinetics.* 75: 25-30.
7. Watkins, R.E., Maglich, J.M., Moore, L.B., Wisely, G.B., Noble, S.M., Davis-Searles, P.R., Lambert, M.H., Kliewer, S.A. and Redinbo, M.R. (2003) A crystal structure of human PXR in complex with the St. John's wort compound hyperforin. *Biochemistry* 42 (6): 1430-1438.

Table 1: Chemical constituents of *Hypericum perforatum* Linn.

Parts used	Chemical composition	Reference
	Red pigment	Hooper, Warden and Dymock
Whole plant	Hypericin	Thomson, 1957
	Dianthrone	Mathis and Ourisson, 1963
	Carotenoid	Costes and Chantal, 1967
	Novoimanin	Derbentseva and Rabinovich 1968
	Tannins	Razinskaite, 1970
	Flavonoids	Maksyutina and Koget. 1971
	Hyperoside	Dittmann *et al.*, 1971

	Hyperforin	Guevich *et al.*, 1971
	Quercitrin	Koget, 1972
	Tannins	Derbentseva *et al.*, 1972
	(+)-catechin	Derbentseva *et al.*, 1972
	Leucocyanidin	Derbentseva *et al.*, 1972
	Terpenes	Sticher, 1977
	Xanthones	Nielsen and Arends, 1978
Oil		Chialva *et al.* 1981
	N-Alkanes	Brondz *et al.*, 1983
	Flavanols	Kitanov, 1983
	Pseudohypericin	Freytag, 1984
	Phenolic acids	Olivier *et al.*, 1985
	Flavonoids	Vasil'chenko *et al.*, 1986
	Flavonoids	Stoyanova *et al.*, 1987
Whole plant	Hyperoside, quercetin, isoquercetin, rutin, methyhespericin, iso-quercitrin, quercitrin, I-3/II-8-biapigenin (amentoflavone) and kaempferol.	Berghoefer and Hoelzl, 1987
	Adhyperforin	Maisenbacher and Kovar, 1992
Leaves and stem	Hyperforin	Rucker *et al.*, 1992
Whole plant	Hyperforin	Wagner and Bladt, 1994
	Protohypericin and protopseudohypericin	Bruneton, 1995
	Hydroperoxycariforin	Rucker *et al.*, 1995
	Napthodianthrones	Newall, Anderson and Phillipson, 1996
	Hypericin	Upton, 1997
	Napthodianthrones, flavonoids, phloro-	Nahrstedt and Butterweck, 1998

	glucinols, xanthones, phenolic acids and terpenes	
Whole plant	Napthodianthrones, flavonoids, phloro-glucinols, xanthones and essential oil	Heiligenstein and Guenther 1998
	Acylphloroglucinols	Erdelmeier, 1998
	Hyperforin	Chatterjee *et al.*, 1998
Aerial parts	Phloroglucinol	Vugdelija *et al.,* 2000
	Miquelianin	Butterweck *et al.*, 2000
	Astilbin	Butterweck *et al.,* 2000
Whole plant	Napthodianthrones, acylphloroglucinols and flavonoids	Telekova *et al.*, 2000
Oil	β-caryophyllene	Gudzic, Dordevic, Palic and Stojanovic 2001
Oil	2-methyl-octane	Gudzic, Dordevic, Palic and Stojanovic, 2001
Whole plant	Napthodianthrones and acylphloroglucinols	Tolonen *et al.*, 2002
	Hyperforin	Adam *et al.*, 2002
Leaves	Hypericin	Piovan *et al.*, 2004
	N-1-alkanols	Brondz and Greibrokk (year not mentioned)
	Flavones	Tsitsina (Year not mentioned)

Table 2. Traditional uses of *Hypericum perforatum*

Use	Reference
Vulnerary	Silberman, 1998
Anthelmentic	Hooper, Warden and Dymock
Diuretic	Hooper, Warden and Dymock
Emmenagouge	Hooper, Warden and Dymock

Antidepressant	Rasmussen, 1998
Anxiolytic	Rasmussen, 1998
Antimalarial,anti-inflammatory, analgesic and gastro protective	Hauselhuber, Kleinachimdt and Wedel, 1969
Vulnerary	Wren, 1975
Vulnerary	Stuart, 1979
Vulnerary	Ruddock, 1937
Vulnerary	Culpeper, 1847
Lithontriptic	Hill, 1808
Antidiarrheal and abortifacient	Moerman, 1986
Vulnerary	King, 1876

Table 3. Investigative work on *Hypericum perforatum* extract and isolated constituents

Extract/ phytochemical	Study	Reference
Imanine	Antimicrobial	Negrash *et al.*, 1969
Flavonoids	Antiviral	Mishenkova *et al.*, 1975
Catechins	Antiviral	Mishenkova *et al.*, 1975
Extract	Anti-inflammatory	Shipochliev *et al.*, 1981.
Hot water extract	Antifungal	Khosla and Bhatia, 1982
Volatile oil	Antifungal	Khosla and Bhatia, 1982
Pseudohypericin	Antiviral.	Takahashi *et al.*, 1989.
Amentoflavone	Anti-ulcerogenic	Berghofer and Holzl, 1989
Amentoflavone	Anti-inflammatory	Berghofer and Holzl, 1989
Procyanidin fraction	Cardio protective	Melzer *et al.*, 1991
Tincture	Vulnerary	Rao *et al.*, 1991
Aqueous extract	Radio protective	Smyshliaeva *et al.*, 1992
Extract	Hepatoprotective	Ozturk *et al.*, 1992
Extract	Antidepressant	Schmidt and Sommer, 1993
Infusion	Anti-ulcerogenic	Krylov and Ibatov, 1993

Extract	Antidepressant	Sommer and Harrer, 1994.
Extract	Antidepressant	Muller and Rossol, 1994
Extract	Antidepressant	Thiele, Brink and Ploch, 1994.
Six fractions	Antidepressant	Bladt and Wagner, 1994
Extract	Antidepressant	Thiede and Walper, 1994
Extract	Antidepressant	Vorbach, Hubner and Arnoldt 1994
Extract	Antidepressant	Hansgen, Vesper and Ploch 1994
Extract	Antidepressant	Hubner, Lande and Podzuweit 1994
Methanolic extract	Antidepressant	Perovic and Muller, 1995
Extract	Antidepressant	Witte *et al.,* 1995
Polyphenol fraction	Immunotropic	Evstifeeva and Sibiriak, 1996
Lipophilic fraction	Immunotropic	Evstifeeva and Sibiriak, 1996
Extract	Analgesic	Ozturk *et al.*, 1996
Extract	Antidepressant	Laakmann, Schule, Baghai and Kieser, 1997
Extract	Antidepressant	Butterweck *et al.*, 1998
Extract	Antidepressant	Wonnermann, Singer and Muller, 1999.
Extract	Anti-inflammatory	Bork *et al.*, 1999
Extract	Antidepressant	Grube *et al.,* 1999
Extract	Antiviral	Koytchev, Alken, and Dundarov, 1999
50%ethanolic extract	Anxiolytic	Kumar, Jaiswal, Singh and Bhattacharya, 2000
50% ethanolic extract	Nootropic	Kumar, Singh, Muruganandam and Bhattacharya, 2000
Xanthones	Antidepressant	Muruganandam *et al.*, 2000
Flavonoids	Antidepressant	Butterweck *et al.*, 2000

Extract	Antidepressant	Kumar, Singh and Bhattacharya, 2001
Extract	Cognitive activity	Klusa *et al.,* 2001
Pseudohypericin	Antidepressant	Simmen *et al.*, 2003
Methanolic extract	Anticancer (prostate)	Martarelli *et al.*, 2004

Table 4. Investigative work on Hypericin

Activity	Reference
Antiviral	Takahashi *et al.*, 1989
Photocytotoxicity	Thomas, 1992
Antifungal	Katz and Donodio, 1993
Antitumor	Katz and Donodio, 1993
Antiviral	Mcauliffe *et al.*, 1993
Antiviral	Klose and Wernet, 1993
Anticancer	Couldwell *et al.*, 1994